PLATE I.   MODELS AND MIMICS IN AUSTRALIAN (*Figs.* 1–3) AND
TROPICAL AMERICAN (*Figs.* 4–7) INSECTS

THE
GENETICAL THEORY OF
NATURAL SELECTION

BY

R. A. FISHER, Sc.D., F.R.S.

DOVER PUBLICATIONS, INC.

NEW YORK

Published in Canada by General Publishing Company, Ltd., 30 Lesmill Road, Don Mills, Toronto, Ontario.
Published in the United Kingdom by Constable and Company, Ltd., 10   Orange Street, London WC 2.

This Dover edition, first published in 1958, is a revised and enlarged version of the work originally published in 1930 by the Oxford University Press.

International Standard Book Number: 0-486-60466-7
Library of Congress Catalog Card Number: 58-13362

Manufactured in the United States of America
Dover Publications, Inc.
180 Varick Street
New York, N. Y. 10014

TO

MAJOR LEONARD DARWIN

*In gratitude for the encouragement,
given to the author, during the last
fifteen years, by discussing many
of the problems dealt with
in this book*

BY THE SAME AUTHOR

Statistical Methods for Research Workers.
1925. Twelfth edition, 1954.
The Design of Experiments. 1935. Sixth edition, (reprinted) 1953.
The Theory of Inbreeding. 1949.
Statistical Methods and Scientific Inference.
1956.
All published by Oliver & Boyd, Edinburgh.

PREFACE

NATURAL Selection is not Evolution. Yet, ever since the two words have been in common use, the theory of Natural Selection has been employed as a convenient abbreviation for the theory of Evolution by means of Natural Selection, put forward by Darwin and Wallace. This has had the unfortunate consequence that the theory of Natural Selection itself has scarcely ever, if ever, received separate consideration. To draw a physical analogy, the laws of conduction of heat in solids might be deduced from the principles of statistical mechanics, yet it would have been an unfortunate limitation, involving probably a great deal of confusion, if statistical mechanics had only received consideration in connexion with the conduction of heat. In this case it is clear that the particular physical phenomena examined are of little theoretical interest compared to the principle by which they can be elucidated. The overwhelming importance of evolution to the biological sciences partly explains why the theory of Natural Selection should have been so fully identified with its role as an evolutionary agency, as to have suffered neglect as an independent principle worthy of scientific study.

The other biological theories which have been put forward, either as auxiliaries, or as the sole means of organic evolution, are not quite in the same position. For advocates of Natural Selection have not failed to point out, what was evidently the chief attraction of the theory to Darwin and Wallace, that it proposes to give an account of the means of modification in the organic world by reference only to 'known', or independently demonstrable, causes. The alternative theories of modification rely, avowedly, on hypothetical properties of living matter which are inferred from the facts of evolution themselves. Yet, although this distinction has often been made clear, its logical cogency could never be fully developed in the absence of a separate investigation of the independently demonstrable modes of causation which are claimed as its basis. The present book, with all the limitations of a first attempt, is at least an attempt to consider the theory of Natural Selection on its own merits.

When the theory was first put forward, by far the vaguest element in its composition was the principle of inheritance. No man of learning or experience could deny this principle, yet, at the time, no approach could be given to an exact account of its working. That an

independent study of Natural Selection is now possible is principally
due to the great advance which our generation has seen in the science
of genetics. It deserves notice that the first decisive experiments,
which opened out in biology this field of exact study, were due to
a young mathematician, Gregor Mendel, whose statistical interests
extended to the physical and biological sciences. It is well known
that his experiments were ignored, to his intense disappointment,
and it is to be presumed that they were never brought under the
notice of any man whose training qualified him to appreciate their
importance. It is no less remarkable that when, in 1900, the genetic
facts had been rediscovered by De Vries, Tschermak, and Correns, and
the importance of Mendel's work was at last recognized, the principal
opposition should have been encountered from the small group of
mathematical statisticians then engaged in the study of heredity.

The types of mind which result from training in mathematics and
in biology certainly differ profoundly; but the difference does not
seem to lie in the intellectual faculty. It would certainly be a mistake
to say that the manipulation of mathematical symbols requires more
intellect than original thought in biology; on the contrary, it seems
much more comparable to the manipulation of the microscope and
its appurtenances of stains and fixatives; whilst original thought in
both spheres represents very similar activities of an identical faculty.
This accords with the view that the intelligence, properly speaking,
is little influenced by the effects of training. What is profoundly
susceptible of training is the imagination, and mathematicians and
biologists seem to differ enormously in the manner in which their
imaginations are employed. Most biologists will probably feel that
this advantage is all on their side. They are introduced early to the
immense variety of living things; their first dissections, even if only
of the frog or dog fish, open up vistas of amazing complexity and
interest, at the time when the mathematician seems to be dealing
only with the barest abstractions, with lines and points, infinitely
thin laminae, and masses concentrated at ideal centres of gravity.
Perhaps I can best make clear that the mathematician's imagination
also has been trained to some advantage, by quoting a remark
dropped casually by Eddington in a recent book—

'We need scarcely add that the contemplation in natural science of a
wider domain than the actual leads to a far better understanding of the
actual.' (p. 267, *The Nature of the Physical World*.)

For a mathematician the statement is almost a truism. From a biologist, speaking of his own subject, it would suggest an extraordinarily wide outlook. No practical biologist interested in sexual reproduction would be led to work out the detailed consequences experienced by organisms having three or more sexes; yet what else should he do if he wishes to understand why the sexes are, in fact, always two? The ordinary mathematical procedure in dealing with any actual problem is, after abstracting what are believed to be the essential elements of the problem, to consider it as one of a system of possibilities infinitely wider than the actual, the essential relations of which may be apprehended by generalized reasoning, and subsumed in general formulae, which may be applied at will to any particular case considered. Even the word possibilities in this statement unduly limits the scope of the practical procedures in which he is trained; for he is early made familiar with the advantages of imaginary solutions, and can most readily think of a wave, or an alternating current, in terms of the square root of minus one. The most serious difficulty to intellectual co-operation would seem to be removed if it were clearly and universally recognized that the essential difference lies, not in intellectual methods, and still less in intellectual ability, but in an enormous and specialized extension of the imaginative faculty, which each has experienced in relation to the needs of his special subject. I can imagine no more beneficial change in scientific education than that which would allow each to appreciate something of the imaginative grandeur of the realms of thought explored by the other.

In the future, the revolutionary effect of Mendelism will be seen to flow from the particulate character of the hereditary elements. On this fact a rational theory of Natural Selection can be based, and it is, therefore, of enormous importance. The merit for this discovery must mainly rest with Mendel, whilst among our countrymen, Bateson played the leading part in its early advocacy. Unfortunately he was unprepared to recognize the mathematical or statistical aspects of biology, and from this and other causes he was not only incapable of framing an evolutionary theory himself, but entirely failed to see how Mendelism supplied the missing parts of the structure first erected by Darwin. His interpretation of Mendelian facts was from the first too exclusively coloured by his earlier belief in the discontinuous origin of specific forms. Though his influence upon

evolutionary theory was thus chiefly retrogressive, the mighty body
of Mendelian researches throughout the world has evidently out-
grown the fallacies with which it was at first fostered. As a pioneer of
genetics he has done more than enough to expiate the rash polemics
of his early writings.

To treat Natural Selection as an agency based independently on
its own foundations is not to mimimize its importance in the theory
of evolution. On the contrary, as soon as we require to form opinions
by other means than by comparison and analogy, such an independent deductive basis becomes a necessity. This necessity is particularly to be noted for mankind ; since we have some knowledge of the
structure of society, of human motives, and of the vital statistics of
this species, the use of the deductive method can supply a more
intimate knowledge of the evolutionary processes than is elsewhere
possible. In addition it will be of importance for our subject to call
attention to several consequences of the principle of Natural Selection
which, since they do not consist in the adaptive modification of specific
forms, have necessarily escaped attention. The genetic phenomena of
dominance and linkage seem to offer examples of this class, the future
investigation of which may add greatly to the scope of our subject.

No efforts of mine could avail to make the book easy reading.
I have endeavoured to assist the reader by giving short summaries
at the ends of all chapters, except Chapter IV, which is summarized
conjointly with Chapter V. Those who prefer to do so may regard
Chapter IV as a mathematical appendix to the corresponding part
of the summary. The deductions respecting Man are strictly inseparable from the more general chapters, but have been placed
together in a group commencing with Chapter VIII. I believe no
one will be surprised that a large number of the points considered
demand a far fuller, more rigorous, and more comprehensive treatment. It seems impossible that full justice should be done to the
subject in this way, until there is built up a tradition of mathematical
work devoted to biological problems, comparable to the researches
upon which a mathematical physicist can draw in the resolution of
special difficulties.

R. A. F.

ROTHAMSTED, *June* 1929.

CONTENTS

DESCRIPTION OF PLATES
PLATE I
Frontispiece

FIG. 1. *Abispa (Monerebia) ephippium,* Fab., a common member of the predominant group of Australian wasps, characterized by a dark brownish-orange ground-colour and the great size and reduced number of the black markings. They are mimicked by many other insects of different groups including bees, flies, moths, and numerous beetles. That the resemblance may be produced by quite different methods is illustrated in Figs. 2 and 3.

FIG. 2. *Tragocerus formosus,* Pascoe, a Longicorn beetle. Almost the whole of the mimetic pattern is developed on the elytra or wing-covers which hide the unwasplike abdomen, shown from above in Fig. 2A. Free movement of the wings is permitted by an arched excavation in the side of each wing-cover.

FIG. 3. *Esthesis ferrugineus,* Macleay. Another Longicorn beetle in which the mimetic pattern is developed on the abdomen itself; the wing-covers are reduced to small rounded scales, thus freeing the wings, but at the same time exposing the abdomen.

FIG. 4. *Heliconius erato erato,* Linn., a distasteful tropical American butterfly with a conspicuous pattern beautifully mimicked by the day-flying Hypsid moth represented below.

FIG. 5. *Pericopis phyleis,* Druce. This moth and its butterfly model were taken in Peru. The striking mimetic resemblance does not extend to the antennae, which are threadlike and inconspicuous in the model.

FIG. 6. *Methona confusa,* Butler, another tropical American butterfly captured with one of its moth mimics (Fig. 7) in Paraguay. In this butterfly and its allies, the antennae are rendered conspicuous by terminal orange knobs, resembled by many of the mimics in different groups of butterflies and moths.

FIG. 7. *Castnia linus,* Cramer. The antennal knobs of this day-flying moth, in spite of the marked resemblance, possess a form quite different from those of the model, but one characteristic of the Castniidae. This example of mimetic likeness to a normally inconspicuous feature, here exceptionally emphasized, may be compared with Figs. 1, 1A–3, 3A on Plate II.

This model and mimic also illustrate, as do Figs. 1–3, the different methods by which the resemblance may be obtained. The pale transparent areas of the model are produced by the great reduction in the size of the scales; in the mimic, without reduction, by their transparency

and by their being set at a different angle so that the light passes between
them. The important mimetic association illustrated by Figs. 6 and 7
includes numerous other species belonging to several distantly related
groups of butterflies and moths, and among these transparency is
attained by various different methods.

PLATE II
Page 175

FIGS. 1–3. The head of the abundant East African Acraeine butterfly
Acraea zetes acara, Hew., as seen from the front (1), from above (2) and
the side (3), showing that the palpi, which are inconspicuous in most
butterflies, are a prominent feature with their orange colour displayed
against the black background.

FIGS. 1A–3A. Similar aspects of the head of the Nymphaline butterfly
Pseudacraea boisduvali trimenii, Butler, a mimic of *A. z. acara* and found
in the same part of Africa. It is evident that the resemblance here extends to the exceptionally emphasized feature, as was observed in the
American examples shown in Figs. 6 and 7 of Plate I.

FIG. 4. *Danaida tytia*, Gray, a conspicuous Oriental Danaine butterfly
taken with its mimic (Fig. 5) in the Darjiling district.

FIG. 5. *Papilio agestor*, Gray, a swallowtail butterfly mimicking the pattern
of *tytia*.

FIG. 6. *Neptis imitans*, Oberth., a Nymphaline butterfly from S. W. China,
mimicking the geographical form of *D. tytia* which is found in the same
area.

Thus these two butterflies of widely separated groups both mimic this
peculiar Danaine pattern.

The butterflies and moths here represented illustrate by single examples
the widespread mimicry of the chief distasteful families in the tropics—on
Plate I the Ithomiinae (Fig. 6) and Heliconinae (Fig. 4) of the New World;
on Plate II, the Danainae (Fig. 5), and Acraeinae (Figs. 1–3) of the Old.

Plate I and Plate II appeared in the first edition in
color. In this edition they are in halftone. The figures
are reduced 8% from their natural size.

THE
GENETICAL THEORY OF
NATURAL SELECTION

I

THE NATURE OF INHERITANCE

The consequences of the blending theory, as drawn by Darwin. Difficulties felt by Darwin. Particulate inheritance. Conservation of the variance. Theories of evolution worked by mutations. Is all inheritance particulate ? Nature and frequency of observed mutations.

But at present, after drawing up a rough copy on this subject, my conclusion is that external conditions do extremely little, except in causing mere variability. This mere variability (causing the child not closely to resemble its parent) I look at as very different from the formation of a marked variety or new species. DARWIN, 1856. (*Life and Letters*, ii, 87.)

As Samuel Butler so truly said : 'To me it seems that the "Origin of Variation", whatever it is, is the only true "Origin of Species".' W. BATESON, 1909.

The consequences of the blending theory

THAT Charles Darwin accepted the fusion or blending theory of inheritance, just as all men accept many of the undisputed beliefs of their time, is universally admitted. That his acceptance of this theory had an important influence on his views respecting variation, and consequently on the views developed by himself and others on the possible causes of organic evolution, was not, I think, apparent to himself, nor is it sufficiently appreciated in our own times. In the course of the present chapter I hope to make clear the logical consequences of the blending theory, and to show their influence, not only on the development of Darwin's views, but on the change of attitude towards these, and other suppositions, necessitated by the acceptance of the opposite theory of particulate inheritance.

It is of interest that the need for an alternative to blending inheritance was certainly felt by Darwin, though probably he never worked out a distinct idea of a particulate theory. In a letter to Huxley probably dated in 1857 occur the sentences (*More Letters*, vol. i, Letter 57).

Approaching the subject from the side which attracts me most, viz., inheritance, I have lately been inclined to speculate, very crudely and indistinctly, that propagation by true fertilization will turn out to be a sort of mixture, and not true fusion, of two distinct individuals, or rather of innumerable individuals, as each parent has its parents and ancestors. I can understand on no other view the way in which crossed forms go back to so large an extent to ancestral forms. But all this, of course, is infinitely crude.

The idea apparently was never developed, perhaps owing to the rush of work which preceded and followed the publication of the *Origin*. Certainly he did not perceive that the arguments on variation in his rough essays of 1842 and 1844, which a year later (1858) he would be rewriting in the form of the first chapter of the *Origin*, required, on a particulate theory, a complete reformulation. The same views indeed are but little changed when 'The Causes of Variability' came to be discussed in Chapter XXII of *Variation of Animals and Plants* published in 1868.

The argument which can be reconstructed from these four sources may be summarized as follows:

(*a*) with blending inheritance bisexual reproduction will tend rapidly to produce uniformity; therefore

(*b*) if variability persists, causes of new variation must be continually at work; hence

(*c*) the causes of the great variability of domesticated species, of all kinds and in all countries, must be sought for in the conditions of domestication; but

(*d*) the only characteristics of domestication sufficiently general to cover all cases are changed conditions and increase of food;

(*e*) some changes of conditions seem to produce definite and regular effects, e. g. increased food causes (hereditary) increase in size, yet the important effect is an indefinite variability in all directions, ascribable to a disturbance, by change of conditions, of the regularity of action of the reproductive system;

(*f*) wild species also will occasionally, by geological changes, suffer changed conditions, and occasionally also a temporary increase in the supply of food; they will therefore, though perhaps rarely, be caused to vary. If on these occasions no selection is exerted the variations will neutralize one another by bisexual reproduction and die away, but if selection is acting, the variations in the right direction will be accumulated and a permanent evolutionary change effected.

To modern readers this will seem a very strange argument with which to introduce the case for Natural Selection; all that is gained by it is the inference that wild as well as domesticated species will at least occasionally present heritable variability. Yet it is used to introduce the subject in the two essays and in the *Origin*. It should be remembered that, at the time of the essays, Darwin had little direct evidence on this point, which, since the power of human selection to modify domesticated races was widely admitted, was a cardinal point in the original argument. Even in the *Origin* the second chapter on 'Variation under Nature' deals chiefly with natural varieties sufficiently distinct to be listed by botanists, and these were certainly regarded by Darwin not as the materials but as the products of evolution. During the twenty-six years between 1842 and 1868 evidence must have flowed in sufficiently at least to convince him that heritable variability was as widespread, though not nearly so great, in wild as in domesticated species. The line of reasoning in question seems to have lost its importance sufficiently for him to introduce the subject in 1868 (*Variation*, Chapter XXII) with the words 'The subject is an obscure one; but it may be useful to probe our ignorance.'

It is the great charm of the essays that they show the *reasons* which led Darwin to his conclusions, whereas the later works often only give the *evidence* upon which the reader is to judge of their truth. The antithesis is not an unnatural one, for every active mind must form opinions without direct evidence, else the evidence too often would never be collected. Impartiality and scientific discipline come into action effectively in submitting the opinions formed to as much relevant evidence as can be made available. The earlier steps in the argument set out above appear only in the two essays, while the conclusions continue almost unchanged up to the *Variation of Animals and Plants*. Indeed the first step (a), logically the most important of all, appears explicitly only in 1842. In 1844 it is clearly implied by its necessary consequences. I believe its significance for the argument of the *Origin* would scarcely ever be detected from a study only of that book. The passage in the 1842 MS. is (*Foundations*, p. 2):

Each parent transmits its peculiarities, therefore if varieties allowed freely to cross, except by the *chance* of two characterized by same peculiarity happening to marry, such varieties will be constantly demolished. All bisexual animals must cross, hermaphrodite plants do cross, it seems very possible that hermaphrodite animals do cross— conclusion strengthened:

together with a partly illegible passage of uncertain position,

If individuals of two widely different varieties be allowed to cross, a third race will be formed—a most fertile source of the variation in domesticated animals. If freely allowed, the characters of pure parents will be lost, number of races thus [illegible] but differences [?] besides the [illegible]. But if varieties differing in very slight respects be allowed to cross, such small variation will be destroyed, at least to our senses— a variation just to be distinguished by long legs will have offspring not to be so distinguished. Free crossing is a great agent in producing uniformity in any breed.

The proposition is an important one, marking as it does the great contrast between the blending and the particulate theories of inheritance. The following proof expresses it in biometrical terms.

Let x and y represent the deviations in any measurement of the two parents from the specific mean; if the measurement is affected not only by inheritance, but by non-heritable (environmental) factors also, x and y stand for the heritable part of these deviations. The amount of heritable variability present in any generation of individuals will be measured by the variance, defined as the mean value of the square of x, or of y. In purely blending inheritance the heritable portions of the deviations of the offspring will be, apart from mutations, equal to $\frac{1}{2}(x + y)$; in the absence of such mutations, therefore, the variance of the progeny generation will be the mean value of $\frac{1}{4}(x^2 + 2xy + y^2)$.

The mean values of x and y are both zero, since they are both defined as deviations from the mean of the species; consequently, in the absence of selective mating, the mean value of xy is also zero, and the variance of the progeny generation is found to be exactly half the variance of the parental generation. More generally the ratio is not $\frac{1}{2}$ but $\frac{1}{2}(1 + r)$, where r is the correlation between x and y. r cannot exceed unity, else the average value of the positive quantities $(x-y)^2$ would have to be negative, and can only be unity, if they are all zero, that is, if the size of each individual prescribes exactly the size of its possible mates. Darwin's 'except by the chance of two individuals characterized by same peculiarities happening to marry' is his way of rejecting high correlations as improbable.

The effect of correlation between mates is to hasten, if the correlation is negative, or to retard if positive, the tendency of blending inheritance to reduce the variance; such effects are not of importance, for even if the correlation were as high as 0·5, and mates had to be as much alike as parent and child usually are, the rate of decay would be little more than halved. The important consequence of the blending is that, if not safeguarded by intense marital correlation, the heritable variance is approximately halved in every generation. To maintain a stationary variance fresh mutations must be available in each generation to supply the half of the variance so lost. If variability persists, as Darwin rightly inferred, causes of new variability must continually be at work. Almost every individual of each generation must be a mutant, i. e. must be influenced by such causes, and moreover must be a mutant in many different characters.

An inevitable inference of the blending theory is that the bulk of the heritable variance present at any moment is of extremely recent origin. One half is new in each generation, and of the remainder one half is only one generation older, and so on. Less than one-thousandth of the variance can be ten generations old; even if by reason of selective mating we ought to say twenty generations, the general conclusion is the same; the variability of domesticated species must be ascribed by any adherent of the blending theory to the conditions of domestication much as they now exist. If variation is to be used by the human breeder, or by natural selection, it must be snapped up at once, soon after the mutation has appeared, and before it has had time to die away. The following passage from the 1844 essay shows that Darwin was perfectly clear on this point (pp. 84–6).

Let us then suppose that an organism by some chance (which might be hardly repeated in 1,000 years) arrives at a modern volcanic island in process of formation and not fully stocked with the most appropriate organisms; the new organism might readily gain a footing, although the external conditions were considerably different from its native ones. The effect of this we might expect would influence in some small degree the size, colour, nature of covering, &c., and from inexplicable influences even special parts and organs of the body. But we might further (and this is far more important) expect that the reproductive system would be affected, as under domesticity, and the structure of the offspring rendered in some degree plastic. Hence almost every part of the body would tend to vary from the typical form in slight degrees, and in no determinate way, and therefore *without selection* the free crossing of these small variations (together with the tendency to reversion to the original form) would constantly be counteracting this unsettling effect of the extraneous conditions on the reproductive system. Such, I conceive, would be the unimportant result without selection. And here I must observe that the foregoing remarks are equally applicable to that small and admitted amount of variation which has been observed in some organisms in a state of nature; as well as to the above hypothetical variation consequent on changes of condition.

Let us now suppose a Being with penetration sufficient to perceive differences in the outer and innermost organization quite imperceptible to man, and with forethought extending over future centuries to watch with unerring care and select for any object the offspring of an organism produced under the foregoing circumstances; I can see no conceivable reason why he could not form a new race (or several were he to separate the stock of the original organism and work on several islands) adapted to new ends. As we assume his discrimination, and his forethought, and his steadiness of object, to be incomparably greater than those qualities in man, so we may suppose the beauty and complications of the adaptations of the new races and their differences from the original stock to be greater than in the domestic races produced by man's agency: the ground-work of his labours we may aid by supposing that the external conditions of the volcanic island, from its continued emergence, and the occasional introduction of new immigrants, vary; and thus to act on the reproductive system of the organism, on which he is at work, and so keep its organization somewhat plastic. With time enough, such a Being might rationally (without some unknown law opposed him) aim at almost any result.

Difficulties felt by Darwin

The argument based on blending inheritance and its logical consequences, though it certainly represents the general trend of Darwin's thought upon inheritance and variation, for some years after he commenced pondering on the theory of Natural Selection, did not satisfy him completely. Reversion he recognized as a fact which stood outside his scheme of inheritance, and that he was not altogether satisfied to regard it as an independent principle is shown by his letter to Huxley already quoted. By 1857 he was in fact on the verge of devising a scheme of inheritance which should include reversion as one of its consequences. The variability of domesticated races, too, presented a difficulty which, characteristically, did not escape him. He notes (pp. 77, 78, *Foundations*) in 1844 that the most anciently domesticated animals and plants are not less variable, but, if anything more so, than those more recently domesticated; and argues that since the supply of food could not have been becoming much more abundant progressively at all stages of a long history of

domestication, this factor cannot alone account for the great variability which still persists. The passage runs as follows:

> If it be an excess of food, compared with that which the being obtained
> in its natural state, the effects continue for an improbably long time;
> during how many ages has wheat been cultivated, and cattle and sheep
> reclaimed, and we cannot suppose their *amount* of food has gone on
> increasing, nevertheless these are amongst the most variable of our
> domestic productions.

This difficulty offers itself also to the second supposed cause of
variability, namely changed conditions, though here it may be
argued that the conditions of cultivation or nurture of domesticated
species have always been changing more or less rapidly. From a
passage in the *Variation of Animals and Plants* (p. 301), which runs:

> Moreover, it does not appear that a change of climate, whether more
> or less genial, is one of the most potent causes of variability; for in
> regard to plants Alph. De Candolle, in his *Geographie Botanique*, re-
> peatedly shows that the native country of a plant, where in most cases
> it has been longest cultivated, is that where it has yielded the greatest
> number of varieties.

it appears that Darwin satisfied himself that the countries in which
animals or plants were first domesticated, were at least as prolific
of new varieties as the countries into which they had been imported,
and it is natural to presume that his inquiries under this head were
in search of evidence bearing upon the effects of changed conditions.
It is not clear that this difficulty was ever completely resolved in
Darwin's mind, but it is clear from many passages that he saw the
necessity of supplementing the original argument by postulating
that the causes of variation which act upon the reproductive system
must be capable of acting in a delayed and cumulative manner so
that variation might still be continued for many subsequent generations.

Particulate inheritance

It is a remarkable fact that had any thinker in the middle of the
nineteenth century undertaken, as a piece of abstract and theoretical
analysis, the task of constructing a particulate theory of inheritance,
he would have been led, on the basis of a few very simple assumptions, to produce a system identical with the modern scheme of
Mendelian or factorial inheritance. The admitted non-inheritance of

scars and mutilations would have prepared him to conceive of the hereditary nature of an organism as something none the less definite because possibly represented inexactly by its visible appearance. Had he assumed that this hereditary nature was completely determined by the aggregate of the hereditary particles (genes), which enter into its composition, and at the same time assumed that organisms of certain possible types of hereditary composition were capable of breeding true, he would certainly have inferred that each organism must receive a definite portion of its genes from each parent, and that consequently it must transmit only a corresponding portion to each of its offspring. The simplification that, apart from sex and possibly other characters related in their inheritance to sex, the contributions of the two parents were equal, would not have been confidently assumed without the evidence of reciprocal crosses; but our imaginary theorist, having won so far, would scarcely have failed to imagine a conceptual framework in which each gene had its proper place or locus, which could be occupied alternatively, had the parentage been different, by a gene of a different kind. Those organisms (homozygotes) which received like genes, in any pair of corresponding loci, from their two parents, would necessarily hand on genes of this kind to all of their offspring alike; whereas those (heterozygotes) which received from their two parents genes of different kinds, and would be, in respect of the locus in question, crossbred, would have, in respect of any particular offspring, an equal chance of transmitting either kind. The heterozygote when mated to either kind of homozygote would produce both heterozygotes and homozygotes in a ratio which, with increasing numbers of offspring, must tend to equality, while if two heterozygotes were mated, each homozygous form would be expected to appear in a quarter of the offspring, the remaining half being heterozygous. It thus appears that, apart from dominance and linkage, including sex linkage, all the main characteristics of the Mendelian system flow from assumptions of particulate inheritance of the simplest character, and could have been deduced *a priori* had any one conceived it possible that the laws of inheritance could really be simple and definite.

In the interval since the foregoing paragraph was written for the 1930 edition of this book, I have had occasion to make (1936, 'Has Mendel's work been rediscovered?' *Annals of Science*, i 115–37) a somewhat detailed study of the structure of the original experiments with peas reported by Mendel in 1865. Many features of the successive decisions he made in conducting these experiments seem unintelligible save on the view that he knew very well what he had to expect, and that the experiments were in reality a confirmation, or demonstration, of a theory at which he had already arrived, perhaps by an equally abstract approach.

In addition to exhibiting in this paper the clear segregation of single and multiple pairs of genes, Mendel also demonstrated in his material the fact of dominance, namely that the heterozygote was not intermediate in appearance, but was almost or quite indistinguishable from one of the homozygous forms. The fact of dominance, though of the greatest theoretical interest, is not an essential feature of the factorial system, and in several important cases is lacking altogether. Mendel also demonstrated what a theorist could scarcely have ventured to postulate, that the different factors examined by him in combination, segregated in the simplest possible manner, namely independently. It was not till after the rediscovery of Mendel's laws at the end of the century that cases of linkage were discovered, in which, for factors in the same linkage group, the pair of genes received from the same parent are more often than not handed on together to the same child. The conceptual framework of loci must therefore be conceived as made of several parts, and these are now identified, on evidence that appears to be singularly complete, with the dark-staining bodies or chromosomes which are to be seen in the nuclei of cells at certain stages of cell division.

The mechanism of particulate inheritance is evidently suitable for producing the phenomenon of reversion, in which an individual resembles a grandparent or more remote ancestor, in some respect in which it differs from its parents; for the ancestral gene combination may by chance be reproduced. This takes its simplest form when dominance occurs, for every union of two heterozygotes will then produce among the offspring some recessives, differing in appearance from their parents, but probably resembling some grandparent or ancestor.

Conservation of the variance

It has not been so clearly recognized that particulate inheritance differs from the blending theory in an even more important fact. There is no inherent tendency for the variability to diminish. In a population breeding at random in which two alternative genes of any factor, exist in the ratio p to q, the three genotypes will occur in the ratio $p^2 : 2pq : q^2$, and thus ensure that their characteristics will be represented in fixed proportions of the population, however they may be combined with characteristics determined by other factors, provided that the ratio $p : q$ remains unchanged. This ratio will indeed be liable to slight changes; first by the chance survival and reproduction of individuals of the different kinds; and secondly by selective survival, by reason of the fact that the genotypes are probably unequally fitted, at least to a slight extent, to their task of survival and reproduction. The effect of chance survival is easily susceptible of calculation, and it appears, as will be demonstrated more fully (Chapter IV), that in a population of n individuals breeding at random the variance will be halved by this cause acting alone in $1\cdot4\ n$ generations. Since the number of individuals surviving to reproduce in each generation must in most species exceed a million, and in many is at least a million-fold greater, it will be seen that this cause of the diminution of hereditary variance is exceedingly minute, when compared to the rate of halving in one or two generations by blending inheritance.

The circumstance that smaller numbers, even less than 100, are sometimes found to reproduce themselves locally, does not, as has been supposed, add to the frequency of random extinction, or to the importance of the so-called 'genetic drift'. For this, perfect isolation is required over a number of generations equally numerous with the population isolated. Even if perfect isolation could be postulated, which is always questionable, it is still improbable that the small isolated population would not ordinarily die out altogether before a period of evolutionary significance could elapse, or that it would not be later absorbed in other populations with a different genetic constitution.

It will be seen in Chapter IV that selection is a much more important agency in keeping the variability of species within limits. But even relatively intense selection will change the ratio $p : q$ of the gene frequencies relatively slowly, and no reasonable assumptions could be made by which the diminution of variance due to selection, in the total absence of mutations, would be much more than a ten-thousandth of that ascribable to blending inheritance. The immediate consequence of this enormous contrast is that the mutation rate needed to maintain a given amount of variability is, on the particulate theory, many thousand times smaller than that which is required on the blending theory. Theories, therefore, which ascribe to agencies believed to be capable of producing mutations, as was 'use and disuse' by Darwin, a power of governing the direction in which evolution is taking place, appear in very different lights, according as one theory of inheritance, or the other, is accepted. For any evolutionary tendency which is supposed to act by favouring mutations in one direction rather than another, and a number of such mechanisms have from time to time been imagined, will lose its force many thousand-fold, when the particulate theory of inheritance, in any form, is accepted; whereas the directing power of Natural Selection, depending as it does on the amount of heritable variance maintained, is totally uninfluenced by any such change. This consideration, which applies to all such theories alike, is independent of the fact that a great part of the reason, at least to Darwin, for ascribing to the environment any considerable influence in the production of mutations, is swept away when we are no longer forced to consider the great variability of domestic species as due to the comparatively recent influence of their artificial environment.

The striking fact, of which Darwin was well aware, that whole brothers and sisters, whose parentage, and consequently whose entire ancestry is identical, may differ greatly in their hereditary composition, bears under the two theories two very different interpretations. Under the blending theory it is clear evidence of new and frequent mutations, governed, as the greater resemblance of twins suggests, by temporary conditions acting during conception and gestation. On the particulate theory it is a necessary consequence of the fact that for every factor a considerable fraction, often not much less than one half, of the population will be heterozygotes, any two offspring of which will be equally likely to receive unlike as like genes from their parents. In view of the close analogy between the statistical concept of variance and the physical concept of energy, we may usefully think of the heterozygote as possessing variance in a potential or latent form, so that instead of being lost it is merely stored in a form from which it will reappear when the heterozygous genotypes are mated. A population mating at random immediately establishes the condition of statistical equilibrium between the latent and the apparent form of variance. The particulate theory of inheritance resembles the kinetic theory of gases with its perfectly elastic collisions, whereas the blending theory resembles a theory of gases with inelastic collisions, and in which some outside agency would be required to be continually at work to keep the particles astir.

The property of the particulate theory of conserving the variance for an indefinite period explains at once the delayed or cumulative effect of domestication in increasing the variance of domesticated species, to which Darwin calls attention. Many of our domesticated varieties are evidently ill-fitted to survive in the wild condition. The mutations by which they arose may have been occurring for an indefinite period prior to domestication without establishing themselves, or appreciably affecting the variance of the wild species. In domestication, however, not only is the rigour of Natural Selection relaxed so that mutant types can survive, and each such survival add something to the store of heritable variance, but novelties of form or colour, even if semi-monstrous, do undoubtedly attract human attention and interest, and are valued by man for their peculiarity. The rapidity with which new variance is accumulated will thus be enhanced. Without postulating any change in the mutation rates due to domestication, we should necessarily infer from what is known of the conditions of domestication that the variation of domesticated species should be greater than that of similar wild species, and that this contrast should be greatest with those species most anciently domesticated. Thus one of the main difficulties felt by Darwin is resolved by the particulate theory.

Theories of evolution worked by mutations

The theories of evolution which rely upon hypothetical agencies, capable of modifying the frequency or direction in which mutations are taking place, fall into four classes. In stating these it will be convenient to use the term 'mutation', to which many meanings have at different times been assigned, to denote simply the initiation of any heritable novelty.

(A) It may be supposed, as by Lamarck in the case of animals, that the mental state, and especially the desires of the organism, possess the power of producing mutations of such a kind, that these desires may be more readily gratified in the descendants. This view postulates (i) that there exists a mechanism by which mutations are caused, and even designed, in accordance with the condition of the nervous system, and (ii) that the desires of animals in general are such that their realization will improve the aptitude of the species for life in its natural surroundings, and also will maintain or improve the aptitude of its parts to co-operate with one another, both in maintaining the vital activity of the adult animal, and in ensuring its normal embryological development. The desires of animals must, in fact, be very wisely directed, as well as being effective in provoking suitable mutations.

(B) A power of adaptation may be widely observed, both among plants and animals, by which particular organs, such as muscles or glands, respond by increased activity and increased size, when additional physiological calls are made upon them. It may be suggested, as it was by Darwin, that such responses of increased functional activity induce, or are accompanied by, mutations of a kind tending to increase the size or activity of the organ in question in future generations, even if no additional calls were made upon this organ's activity. This view implies (i) that the power which parts of organisms possess, of responding adaptively to increased demands upon them, is not itself a product of evolution, but is postulated as a primordial property of living matter: and requires (ii) that a mechanism exists by which the adaptive response shall itself tend to cause, or to be accompanied by, an appropriate mutation.

Both these two suggested means of evolution expressly aim at explaining, not merely the progressive change of organic beings, but the aptitude of the organism to its place in nature, and of its parts to their function in the organism.

(C) It may be supposed that the environment in which the organism is placed controls the nature of the mutations which occur in it, and so directs its evolutionary course; much as the course of a projectile is controlled by the field of force in which it flies.

(D) It may be supposed that the mutations which an organism undergoes are due to an 'inner urge' (not necessarily connected with its mental state) implanted in its primordial ancestors, which thereby directs its predestined evolution.

The two last suggestions give no particular assistance towards the understanding of adaptation, but each contains at least this element of truth; that however profound our ignorance of the causes of mutation may be, we cannot but ascribe them, within the order of Nature as we know it, either to the nature of the organism, or to that of its surrounding environment, or, more generally, to the interaction of the two. What is common, however, to all four of these suppositions, is that each one postulates that the direction of evolutionary change is governed by the predominant direction in which mutations are taking place. However reasonable such an assumption might have seemed when, under the blending theory of inheritance, every individual was regarded as a mutant, and probably a multiple mutant, it is impossible to let it pass unquestioned, in face of the much lower mutation rates appropriate to the particulate theory.

A further hypothetical mechanism, guiding the evolution of the species according to the direction in which mutations are occurring, was suggested by Weismann. Weismann appreciated much more thoroughly than many of his contemporaries the efficacy of Natural Selection, in promoting the adaptation of organisms to the needs of their lives in their actual habitats. He felt, however, that this action would be aided in a subordinate degree if the process of mutation could acquire a kind of momentum, so that a series of mutations affecting the increase or decrease of a part should continue to occur, as a consequence of an initial chance tendency towards such increase or decrease. Such an assumed momentum in the process of mutation he found useful in two respects: (i) it would enable an assumed minimal mutation in an advantageous direction to be increased by further mutations, until it 'attains selection value'; (ii) it explains the continuous decrease of a useless organ, without assuming that each step of this decrease confers any advantage upon the organism manifesting it.

The concept of attaining selection value, which is fairly common in biological literature, seems to cover two distinct cases. In the first case we may imagine that, with increasing size, the utility of an organ shows no increase up to a certain point, but that beyond this point increasing size is associated with increasing utility. In such a case, which, in view of the actual variability of every organism, and of the parts of related organisms, must be regarded as somewhat ideal, we are really only concerned with the question whether the actual variability in different members of the species concerned, does or does not reach as far as the critical point. If it does not do so the species will not be able to take the advantage offered, simply because it is not variable enough, and the postulate of an element of momentum in the occurrence of mutations, was certainly not made in order to allow organisms to be more variable than they would be without it.

The second meaning, which is also common in the literature, depends upon a curious assumption as to the manner in which selective advantage increases with change of size of the organ in question; for it is sometimes assumed that, while at all sizes an increase of size may be advantageous, this advantage increases, not continuously, but in a step-like manner; or at least that increases below a certain limit produce an advantage which may be called 'inappreciable', and therefore neglected. Both the metaphor and the underlying idea appear to be drawn from psychophysical experience. If we compare two physical sensations such as those produced by the weights of two objects, then when the weights are sufficiently nearly equal the subject will often be unable to distinguish between them, and will judge them equal, whereas with a greater disparity, a distinct or appreciable difference of weight is discerned. If, however, the same test is applied to the subject repeatedly with differences between the weights varying from what is easily discernible to very much smaller quantities, it is found that differences in the weights, which would be deemed totally inappreciable, yet make a significant and perfectly regular difference to the *frequency* with which one is judged heavier than the other. The discontinuity lies in our interpretation of the sensations, not in the sensations themselves. Now, survival value is measured by

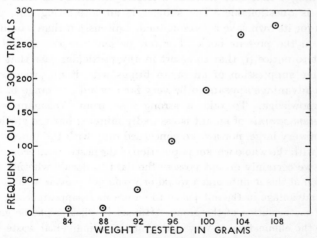

Fig. 1. The frequency with which test objects of different weights are judged heavier than a standard 100 gram weight. (Urban's data, for a single subject.) Illustrating the fact that with a sufficient number of trials, differences in weight, however 'inappreciable', will affect the frequency of the judgement.

the frequency with which certain events, such as death and reproduction, occur, to different sorts of organisms exposed to the different chances of the same environment, and even if we should otherwise be in doubt, the psychophysical experiments make it perfectly clear that the selective advantage will increase or decrease continuously, even for changes much smaller than those appreciable to our own senses, or to those of the predator or other animal, which may enter into the biological situation concerned.   If a change of 1 mm. has selection value, a change of 0·1 mm. will usually have a selection value approximately one-tenth as great, and the effect cannot be ignored because we deem the stimulus inappreciable. The rate at which a mutation increases in numbers at the expense of its allelomorph will indeed depend on the selective advantage it confers, but the rate at which a species responds to selection in favour of any increase or decrease of parts depends on the total heritable variance available, and not on whether this is supplied by large or small mutations. There is no *limen* of appreciable selection value to be considered.

The remaining advantage which Weismann sought in postulating his mechanism of germinal selection was to supply an explanation of the progressive diminution of useless organs, even when these are of so trifling a character that the selective advantage of their suppression is questionable.   The subject is an interesting one, and deserves for its own sake a more extended discussion than would be suitable in the present book.   For our present purpose it will be sufficient to notice (i) that to assert in any particular case that the progressive suppression of an organ brings with it no progressive selective advantage appears to be very far beyond the range of our actual knowledge.   To take a strong case from Weismann—the *receptaculum seminis* of an ant is assuredly minute; but the ant herself is not very large, nor are we concerned only with the individual ant, but with the whole worker population of the nest.   As an economic problem we certainly do not possess the data to decide whether the suppression of this minute organ would or would not provide a sufficient selective advantage in the ant polity to ensure its disappearance in the course of the evolution of the group.   Human parallels might be given in which the elimination of very minute items of individual waste can lend an appreciable support to social institutions which are certainly not negligible.   I do not assert that the suppression of the *receptaculum* has been useful to the ant, but that in this as in other cases, if we pause to give the matter due consideration, it is at once apparent that we have not the knowledge on which to base any decided answer.   (ii) In the second place Weismann's view that in the absence of all selection a useless organ might diminish, degenerate, and finally disappear, by the cumulative action of successive mutations, and especially his view that this is the only type of progressive change which could take place by mutations only, without the guidance of Natural Selection, is fully in accordance with modern knowledge of the nature of mutations.   The special mechanism, however, by which he sought to explain the successive occurrence of degenerative mutations must be judged to be superfluous.   It is moreover exposed to the logical objection that the driving force of his mechanism of germinal selection is an assumed competition for nutriment between the chromatin elements which represent the degenerating organ, and those which represent the rest of the body.   The degenerating organ itself is assumed to be so unimportant that its demands upon the general nutrition of the body are to be neglected; and it may well be asked if it is legitimate to bring in, in respect of the well-nourished germ cell, the factor of nutritional competition which is to be ignored in the occasionally ill-nourished body.

Is all inheritance particulate?

The logical case for rejecting the assumption that the direction of evolutionary change is governed by the direction in which mutations are taking place, and thereby rejecting the whole group of theories in which this assumption is implicit, would be incomplete had not modern researches supplied the answer to two further questions: (i) May it not be that in addition to the mechanism of particulate inheritance, which has been discovered and is being investigated, there is also, in living organisms, an undiscovered mechanism of blending inheritance?   (ii) Do the known facts within the particulate system render a mechanism, which could control the predominant direction of mutation, inoperative as a means of governing the direction of evolutionary change?

On the first point it should be noted briefly that, whereas at the beginning of the century there were several outstanding facts of inheritance which seemed to demand some sort of blending theory, these have all in the course of research been shown, not only to be compatible with particulate inheritance, but to reveal positive indications that such is their nature.   The apparent blending in colour in crosses between white races of man and negroes is compatible with the view that these races differ in several Mendelian factors, affecting the pigmentation. Of these some may have intermediate heterozygotes, and of the remainder in some the darker, and in some the lighter tint may be dominant. The Mendelian theory is alone competent to explain the increased variability of the offspring of the mulattoes.

The biometrical facts as to the inheritance of stature and other human measurements, though at first regarded as incompatible with the Mendelian system, have since been shown to be in complete accordance with it, and to reveal features not easily explicable on any other view. The approximately normal distribution of the measurements themselves may be deduced from the simple supposition that the factors affecting human stature are approximately additive in their effects. The correlations found between relatives of different degrees of kinship are, within their sampling errors, of the magnitudes which would be deduced from the assumption that the measurement is principally determined by inheritance, and that the factors controlling it show, like most Mendelian factors, complete or almost complete dominance. The presence of dominance is a Mendelian feature, which is shown in the biometrical data by the well-established fact that children of the same parents are, on the average, somewhat more alike than are parent and offspring.

So far we have merely established the negative fact that there are no outstanding observations which require a blending system of inheritance. There is, however, one group of modern researches which, at least in the organisms investigated, seems to exclude it, even as a possibility. In certain organisms which are habitually self-fertilized, as Johannsen was the first to show with a species of bean, it is possible to establish so-called pure lines, within which heritable variability is, apart from exceptional mutations, completely absent. Within these lines the selection of the largest or the smallest beans, even where this selection was continued for ten or twenty generations, constantly produced offspring of the same average size. This size differed from one line to another, showing that heritable variability existed abundantly in the species, and among the thousands of beans examined two distinct mutants were reported. If, however, any appreciable fraction of the variance in bean size were ascribable to elements which blend, the mutations necessary to maintain such heritable variability would, in ten generations, have had time to supply it almost to its maximum extent, and must inevitably have been revealed by selection. Experiments of this type seem capable of excluding the possibility that blending inheritance can account for any appreciable fraction of the variance observed.

Nature and frequency of observed mutations

The assumption that the direction of evolutionary change is actually governed by the direction in which mutations are occurring is not easily compatible with the nature of the numerous mutations which have now been observed to occur. For the majority of these produce strikingly disadvantageous deformities, and indeed much the largest class are actually lethal. If we had to admit, as has been so often assumed in theory, that these mutations point the direction of evolution, the evolutionary prospects of the little fruit-fly *Drosophila* would be deplorable indeed. Nor is the position apparently different with man and his domesticated animals and plants ; as may be judged from the frequency with which striking recessive defects, producing, for example, albinism, deaf-mutism, and feebleness of mind in man, must have occurred in the comparatively recent past, as mutations. Mutant defects seem to attack the human eye as much as that of *Drosophila*, and in general the mutants which occur in domesticated races are often monstrous and predominantly defective, whereas we know in many cases that the evolutionary changes which these creatures have undergone under human selection have been in the direction of a manifest improvement.

In addition to the defective mutations, which by their conspicuousness attract attention, we may reasonably suppose that other less obvious mutations are occurring which, at least in certain surroundings, or in certain genetic combinations, might prove themselves to be beneficial. It would be unreasonable, however, to assume that such mutations appear individually with a frequency much greater than that which is observed in the manifest defects. The frequency of individual mutations in *Drosophila* is certainly seldom greater than one in 100,000 individuals, and we may take this figure to illustrate the inefficacy of any agency, which merely controls the predominant direction of mutation, to determine the predominant direction of evolutionary change. For even if selection were totally absent, a lapse of time of the order of 100,000 generations would be required to produce an important* change with respect to the factor

* E.g., from 0 to 63·2% or from 25% to 72·4%.

concerned, in the heritable nature of the species.   Moreover, if the mutant gene were opposed, even by a very minute selective disadvantage, the change would be brought to a standstill at a very early stage.

The ideas necessary for a precise examination of the nature of selective advantage will be developed in Chapter II; but it will be readily understood that if we speak of a selective advantage of one per cent., with the meaning that animals bearing one gene have an expectation of offspring only one per cent. greater than those bearing its allelomorph, the selective advantage in question will be a very minute one; at least in the sense that it would require an enormous number of experimental animals, and extremely precise methods of experimentation, to demonstrate so small an effect experimentally. Such a selective advantage would, however, greatly modify the genetic constitution of the species, not in 100,000 but in 100 generations.*   If, moreover, we imagine these two agencies opposed in their tendencies, so that a mutation which persistently occurs in one in 100,000 individuals, is persistently opposed by a selective advantage of only one per cent., it will easily be seen that an equilibrium will be arrived at when only about one individual in 1,000 of the population will be affected by the mutation.   This equilibrium, moreover, will be stable; for if we imagine that by some chance the number of mutants is raised to a higher proportion than this, the proportion will immediately commence to diminish under the action of selection, and evolution will proceed in the direction contrary to the mutation which is occurring, until the proportion of mutant individuals again reaches its equilibrium value.   For mutations to dominate the trend of evolution it is thus necessary to postulate mutation rates immensely greater than those which are known to occur, and of an order of magnitude which, in general, would be incompatible with particulate inheritance.

Summary

The tacit assumption of the blending theory of inheritance led Darwin, by a perfectly cogent argument, into a series of speculations, respecting the causes of variations, and the possible evolutionary effects of these causes.   In particular the blending theory, by the enormous mutation rates which it requires, led Darwin to some extent and others still more to attach evolutionary importance to hypothetical agencies which control the production of mutations.   A mechanism

* This is a slight exaggeration; still in 200 generations the percentage would change from 26·9% to 73·1%.

(Mendelism) of particulate inheritance has since been discovered, requiring mutations to an extent less by many thousandfold. The 'pure line' experiments seem to exclude blending inheritance even as a subordinate possibility. The nature of the mutations observed is not compatible with the view that evolution is directed by their means, while their observed frequency of occurrence shows that an agency controlling mutations would be totally ineffectual in governing the direction of evolutionary change.

The whole group of theories which ascribe to hypothetical physiological mechanisms, controlling the occurrence of mutations, a power of directing the course of evolution, must be set aside, once the blending theory of inheritance is abandoned. The sole surviving theory is that of Natural Selection, and it would appear impossible to avoid the conclusion that if any evolutionary phenomenon appears to be inexplicable on this theory, it must be accepted at present merely as one of the facts which in the current state of knowledge does seem inexplicable. The investigator who faces this fact, as an unavoidable inference from what is now known of the nature of inheritance, will direct his inquiries confidently towards a study of the selective agencies at work throughout the life history of the group in their native habitats, rather than to speculations on the possible causes which influence their mutations. The experimental study of agencies capable of influencing mutation rates is of the highest interest for the light which it may throw on the nature of these changes. We should altogether misinterpret the value of such researches were we to regard them as revealing the causes of evolutionary modification.

Since the discovery of the means of releasing atomic energy, and the demonstration in 1945 of its immense potency in the destruction of life and property, considerable popular alarm has been promoted as to the possibility of injury to the human race due to mutations caused by various physical radiations arising from its utilization. Wholesale destruction is certainly terrible, but this is no reason for ascribing to the same cause in addition terrible consequences of an entirely hypothetical kind. In respect of human evolution, among the grave dangers to which the human species is exposed, mutations would seem to be among those least to be feared; moreover it is now known that mutagenic action is characteristic not only of high energy radiations, but of a wide variety of purely chemical agents.

II

THE FUNDAMENTAL THEOREM OF NATURAL SELECTION

The life table and the table of reproduction. The Malthusian parameter of population increase. Reproductive value. The genetic element in variance. Natural Selection. The nature of adaptation. Deterioration of the environment. Changes in population. Summary.

One has, however, no business to feel so much surprise at one's ignorance, when one knows how impossible it is without statistics to conjecture the duration of life and percentage of deaths to births in mankind. DARWIN, 1845. (*Life and Letters*, ii, 33.)

In the first place it is said—and I take this point first, because the imputation is too frequently admitted by Physiologists themselves—that Biology differs from the Physico-chemical and Mathematical sciences in being 'inexact'. HUXLEY, 1854.

The life table

IN order to obtain a distinct idea of the application of Natural Selection to all stages in the life-history of an organism, use may be made of the ideas developed in the actuarial study of human mortality. These ideas are not in themselves very recondite, but being associated with the laborious computations and the technical notation employed in the practical business of life insurance, are not so familiar as they might be to the majority of biologists. The textbooks on the subject, moreover, are devoted to the chances of death, and to monetary calculations dependent on these chances, whereas in biological problems at least equal care and precision of ideas is requisite with respect to reproduction, and especially to the combined action of these two agencies in controlling the increase or decrease of the population.

The object of the present chapter is to combine certain ideas derivable from a consideration of the rates of death and reproduction of a population of organisms, with the concepts of the factorial scheme of inheritance, so as to state the principle of Natural Selection in the form of a rigorous mathematical theorem, by which the rate of improvement of any species of organisms in relation to its environment is determined by its present condition.

The fundamental apparatus of the actuary's craft is what is known

22

as a life table. This shows, for each year of age, of the population considered, the proportion of persons born alive who live to attain that age. For example, a life table may show that the proportion of persons living to the age of 20 is 88 per cent., while only 80 per cent. reach the age of 40. It will be easily inferred that 12 per cent. of those born alive die in the first 20 years of life, and 8 per cent. in the second 20 years. The life table is thus equivalent to a statement of the frequency distribution of the age of death in the population concerned. The amount by which each entry is less than the preceding entry represents the number of deaths between these limits of age, and this divided by the number living at the earlier age gives for these the probability of death within a specified time. Since the probability of death changes continuously throughout life, the death rate at a given age can only be measured consistently by taking the age interval to be infinitesimal.   Consequently if l_x is the number living to age x, the death rate at age x is given by:

$$\mu_x = -\frac{1}{l_x}\frac{d}{dx}l_x = -\frac{d}{dx}(\log l_x),$$

the logarithm being taken, as in most mathematical representations, to be on the Natural or Naperian system. The life table thus contains a statement of the death rates at all ages, and conversely can be constructed from a knowledge of the course taken by the death rate throughout life. This in fact is the ordinary means of constructing the life tables in practical use.

It will not be necessary to discuss the technical procedure employed in the construction of life tables, the various conventions employed in this form of statement, nor the difficulties which arise in the interpretation of the observational data available in practice for this purpose. It will be sufficient to state only one point. As in all other experimental determinations of theoretical values, the accuracy attainable in practice is limited by the extent of the observations; the result derived from any finite number of observations will be liable to an error of random sampling, but this fact does not, in any degree, render such concepts as death rates or expectations of life obscure or inexact. These are statements of probabilities, averages &c., pertaining to the hypothetical population sampled, and depend only upon its nature and circumstances. The inexactitude of our methods of measurement has no more reason in statistics than it has in physics to dim our conception of that which we measure. These conceptions would be equally clear if we were stating the chances of death of a single individual of unique genetic constitution, or of one exposed to an altogether transient and exceptional environment.

The table of reproduction

The life table, although itself a very comprehensive statement, is still inadequate to express fully the relation between an organism and its environment; it concerns itself only with the chances or frequency of death, and not at all with reproduction. To repair this deficiency it is necessary to introduce a second table giving rates of reproduction in a manner analogous to the rates of death at each age. Just as a person alive at the beginning of any infinitesimal age interval dx has a chance of dying within that interval measured by $\mu_x dx$, so the chance of reproducing within this interval will be represented by $b_x dx$, in which b_x may be called the rate of reproduction at age x. Again, just as the chance of a person chosen at birth dying within a specified interval of age dx is $l_x \mu_x dx$, so the chance of such a person living to reproduce in that interval will be $l_x b_x dx$.

Owing to bisexual reproduction a convention must be introduced into the measurement of b_x, for each living offspring will be credited to both parents, and it will seem proper to credit each with one half in respect of each offspring produced. This convention will evidently be appropriate for those genes which are not sex-linked (autosomal genes) for with these the chance of entering into the composition of each offspring is known to be one half. In the case of sex-linked genes those of the heterogametic parent will be perpetuated or not according as the offspring is male or female. These sexes, it is true, will not be produced in exactly equal numbers, but since both must co-operate in each act of sexual reproduction, it is clear that the different frequencies at birth must ultimately be compensated by sexual differences in the rates of death and reproduction, with the result that the same convention appears in this case to be equally appropriate.

A similar convention, appropriate in the sense of bringing the formal symbolism of the mathematics into harmony with the biological facts, may be used with respect to the period of gestation. For it will happen occasionally that a child is born after the death of its father. The children born to fathers aged x should in fact be credited to males aged three-quarters of a year younger. Such corrections are not a necessity to an exact mathematical representation of the facts, but are a manifest convenience in simplifying the form of expression; thus with mankind we naturally think of the stage in the life-history as measured in years from birth. With other organisms the variable x which with man represents this age, may in some cases be more conveniently used to indicate rather the stage in the life history irrespective of chronological age, merely to give greater vividness to the meaning of the symbolism, but without altering the content of the symbolical statements.

The Malthusian parameter of population increase

If we combine the two tables giving the rates of death and reproduction, we may, still speaking in terms of human populations, at once calculate the expectation of offspring of the newly-born child. For the expectation of offspring in each element of age dx is $l_x b_x dx$, and the sum of these elements over the whole of life will be the total expectation of offspring. In mathematical terms this is

$$\int_0^\infty l_x b_x dx,$$

where the integral is extended from zero, at birth, to infinity, to cover every possible age at which reproduction might conceivably take place. If at any age reproduction ceases absolutely, b_x will thereafter be zero and so give a terminating integral under the same form.

The expectation of offspring determines whether in the population concerned the reproductive rates are more or less than sufficient to balance the existing death rates. If its value is less than unity the reproductive rates are insufficient to maintain a stationary population, in the sense that any population which constantly maintained the death and reproduction rates in question would, apart from temporary fluctuations, certainly ultimately decline in numbers at a calculable rate. Equally, if it is greater than unity, the population biologically speaking is more than holding its own, although the actual number of heads to be counted may be temporarily decreasing.

This consequence will appear most clearly in its quantitative aspect if we note that corresponding to any system of rates of death and reproduction, there is only one possible constitution of the population in respect of age, which will remain unchanged under the action of this system. For if the age distribution remains unchanged the relative rate of increase or decrease of numbers at all ages must be the same; let us represent the relative rate of increase by m; which will also represent a decrease if m is negative. Then, owing to the constant rates of reproduction, the rate at which births are occurring at any epoch will increase proportionately to e^{mt}. At any particular epoch, for which we may take $t=0$, the rate at which births were occurring x years ago will be proportional to e^{-mx}, and this is the rate at which births were occurring at the time persons now of age x were being born. The number of persons in the infinitesimal age interval dx will therefore be proportional to $e^{-mx}l_x dx$, for of those born only the fraction l_x survive to this age. The age distribution is therefore determinate if the number m is uniquely determined. But knowing the numbers living at each age, and the reproductive rates at each age, the rate at which births are now occurring can be calculated, and this can be equated to the known rate of births appropriate to $t = 0$. In fact, the contribution to the total rate, of persons in the age interval dx, must be $e^{-mx}l_x b_x dx$ and the aggregate for all ages must be

$$\int_0^\infty e^{-mx}l_x b_x dx,$$

which, when equated to unity, supplies an equation for m, of which one and only one real solution exists. Since e^{-mx} is less than unity for all values of x, if m is positive, and is greater than unity for all values of x, if m is negative, it is evident that the value of m, which reduces the integral above expressed to unity, must be positive if the expectation of offspring exceeds unity, and must be negative if it falls short of unity.

The number m which satisfies this equation is thus implicit in any given system of rates of death and reproduction, and measures the relative rate of increase or decrease of a population when in the steady state appropriate to any such system. In view of the emphasis laid by Malthus upon the 'law of geometric increase' m may appropriately be termed the Malthusian parameter of population increase. It evidently supplies in its negative values an equally good measure of population decrease, and so covers cases to which, in respect of mankind, Malthus paid too little attention.

In view of the close analogy between the growth of a population supposed to follow the law of geometric increase, and the growth of capital invested at compound interest, it is worth noting that if we regard the birth of a child as the loaning to him of a life, and the birth of his offspring as a subsequent repayment of the debt, the method by which m is calculated shows that it is equivalent to answering the question—At what rate of interest are the repayments the just equivalent of the loan ? For the unit investment has an expectation of a return $l_x b_x dx$ in the time interval dx, and the present value of this repayment, if m is the rate of interest, is $e^{-mx} l_x b_x dx$; consequently the Malthusian parameter of population increase is the rate of interest at which the present value of the births of offspring to be expected is equal to unity at the date of birth of their parent. The actual values of the parameter of population increase, even in sparsely populated dominions, do not, however, seem to approach in magnitude the rates of interest earned by money, and negative rates of interest are, I suppose, unknown to commerce.

Reproductive value

The analogy with money does, however, make clear the argument for another simple application of the combined death and reproduction rates. We may ask, not only about the newly born, but about persons of any chosen age, what is the present value of their future offspring; and if present value is calculated at the rate determined as before, the question has the definite meaning—To what extent will persons of this age, on the average, contribute to the ancestry of future generations ? The question is one of some interest, since the direct action of Natural Selection must be proportional to this contribution. There will also, no doubt, be indirect effects in cases in which an animal favours or impedes the survival or reproduction of its relatives; as a suckling mother assists the survival of her child, as in mankind a mother past bearing may greatly promote the reproduction of her children, as a foetus and in less measure a sucking child inhibits conception, and most strikingly of all as in the services of neuter insects to their queen. Nevertheless such indirect effects will in very many cases be unimportant compared to the effects of personal reproduction, and by the analogy of compound interest the present value of the future offspring of persons aged x is easily seen to be given by the equation

$$v_x / v_0 = \frac{e^{mx}}{l_x} \int_x^\infty e^{-mt} l_t b_t dt.$$

Each age group may in this way be assigned its appropriate

reproductive value. Fig. 2 shows the reproductive value of women according to age as calculated from the rates of death and reproduction current in the Commonwealth of Australia about 1911. The Malthusian parameter was at that time positive, and as judged from

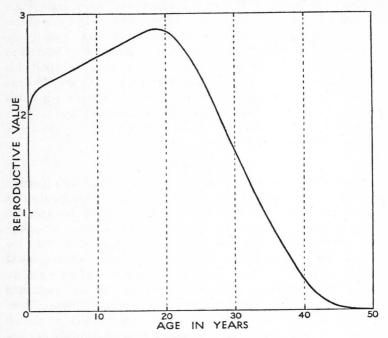

FIG. 2. Reproductive value of Australian women.
The reproductive value for female persons calculated from the birth- and death-rates current in the Commonwealth of Australia about 1911. The Malthusian parameter is +0·01231 per annum.

female rates was nearly equivalent to 1¼ per cent. compound interest; the rate would be lower for the men, and for both sexes taken together, owing to the excess of men in immigration. The reproductive value, which of course is not to be confused with the reproductive rate, reaches its maximum at about 18½, in spite of the delay in reproduction caused by civilized marriage customs; indeed it would have been as early as 16, were it not that a positive rate of interest gives higher value to the immediate prospect of progeny of an older woman, compared to the more remote children of a young girl. If this is the case among a people by no means precocious in reproduction, it would be surprising if, in a state of society entailing marriage at or soon after puberty, the age of maximum reproductive value should fall at any later age than twelve. In the Australian data, the value at birth is lower, partly by reason of the effect of an increasing population in setting a lower value upon remote children and partly because of the risk of death before the reproductive age is reached. The value shown is probably correct, apart from changes in the rate since 1911, for such a purpose as assessing how far it is worth while to give assistance to immigrants in respect of infants (though of course, it takes no account of the factor of eugenic quality), for such infants will usually emigrate with their parents; but it is overvalued from the point of view of Natural Selection to a considerable extent, owing to the capacity of the parents to replace a baby lost during lactation. The reproductive value of an older woman on the contrary is undervalued in so far as her relations profit by her earnings or domestic assistance, and this to a greater extent from the point of view of the Commonwealth, than from that of Natural Selection. It is probably not without significance in this connexion that the death rate in Man takes a course generally inverse to the curve of reproductive value. The minimum of the death rate curve is at twelve, certainly not far from the primitive maximum of the reproductive value; it rises more steeply for infants, and less steeply for the elderly than the curve of reproductive value falls, points which qualitatively we should anticipate, if the incidence of natural death had been to a large extent moulded by the effects of differential survival.

A property that well illustrates the significance of the method of valuation, by which, instead of counting all individuals as of equal value in respect of future population, persons of each age are assigned an appropriate value v_x, is that, whatever may be the age constitution of a population, its total reproductive value will increase or decrease according to the correct Malthusian rate m, whereas counting all heads as equal this is only true in the theoretical case in which the population is in its steady state. For suppose the number of persons in the age interval dx is $n_x dx$; the value of each element of the population will be $n_x v_x dx$; in respect of each such group there will be a gain in value by reproduction at the rate of $n_x b_x v_o dx$, a loss by death of $n_x \mu_x v_x dx$, and a loss by depreciation of $-n_x dv_x$, or in all

$$n_x \{ (b_x v_o - \mu_x v_x)\, dx + dv_x \},$$

but by differentiating the equation by which v_x is defined, it appears that

$$\frac{1}{v_x}\frac{dv_x}{dx} + \frac{1}{l_x}\frac{dl_x}{dx} - m = \frac{-l_x b_x e^{-mx}}{\frac{v_x}{v_o}l_x e^{-mx}} = -\frac{b_x v_o}{v_x},$$

or that

$$dv_x - \mu_x v_x dx + b_x v_o dx = m v_x dx.$$

Consequently the rate of increase in the total value of the population is m times its actual total value, irrespective of its constitution in respect of age. A comparison of the total values of the population at two census epochs thus shows, after allowance for migration, the genuine biological increase or decrease of the population, which may be entirely obscured or reversed by the crude comparison of the number of heads. The population of Great Britain, for example, must have commenced to decrease biologically at some date obscured by the war, between 1911 and 1921, but the census of 1921 showed a nominal increase of some millions, and that of 1931 will, doubtless in less degree, certainly indicate a further spurious period of increase, due to the accumulation of persons at ages at which their reproductive value is negligible.

The genetic element in variance

Let us now consider the manner in which any quantitative individual measurement, such as human stature, may depend upon the individual genetic constitution. We may imagine, in respect of any pair of alternative genes, the population divided into two portions, each comprising one homozygous type together with half of the heterozygotes, which must be divided equally between the two portions. The difference in average stature between these two groups may then be termed the *average excess* (in stature) associated with the gene substitution in question. This difference need not be wholly due to the single gene, by which the groups are distinguished, but possibly also to other genes statistically associated with it, and having similar or opposite effects. This definition will appear the more appropriate if, as is necessary for precision, the population used to determine its value comprises, not merely the whole of a species in any one generation attaining maturity, but is conceived to contain all the genetic combinations possible, with frequencies appropriate to their actual probabilities of occurrence and survival, whatever these may be, and if the average is based upon the statures attained by all these genotypes in all possible environmental circumstances, with frequencies appropriate to the actual probabilities of encountering these circumstances. The statistical concept of the excess in stature of a given gene substitution will then be an exact one, not dependent upon chance as must be any practical estimate of it, but only upon the genetic nature and environmental circumstances of the species. The excess in a factor will usually be influenced by the actual frequency ratio $p : q$ of the alternative genes, and may also be influenced, by way of departures from random mating, by the varying reactions of the factor in question with other factors; it is for this reason that its value for the purpose of our argument is defined in the precise statistical manner chosen, rather than in terms of the average sizes of pure genotypes, as would be appropriate in specifying such a value in an experimental population, in which mating is under control, and in which the numbers of the different genotypes examined is at the choice of the experimenter.

For the same reasons it is also necessary to give a statistical definition of a second quantity, which may be easily confused with that just defined, and may often have a nearly equal value, yet which must be distinguished from it in an accurate argument; namely the *average effect* produced in the population as genetically constituted, by the substitution of the one type of gene for the other. By whatever rules mating, and consequently the frequency of different gene combinations, may be governed, the substitution of a small proportion of the genes of one kind by the genes of another will produce a definite proportional effect upon the average stature. The amount of the difference produced, on the average, in the total stature of the population, for each such gene substitution, may be termed the average effect of such substitution, in contra-distinction to the average excess as defined above. In human stature, for example, the correlation found between married persons is sufficient to ensure that each gene tending to increase the stature must be associated with other genes having a like effect, to an extent sufficient to make the average excess associated with each gene substitution exceed its average effect by about a quarter.

If a is the magnitude of the *average excess* of any factor, and α the magnitude of the *average effect* on the chosen measurement, we shall now show that the contribution of that factor to the genetic variance is represented by the expression $2pqa\alpha$.

The variable measurement will be represented by x, and the relation of the quantities a to it may be made more clear by supposing that for any specific gene constitution we build up an 'expected' value, X, by adding together appropriate increments, positive or negative, according to the natures of the genes present. This expected value will not necessarily represent the real stature, though it may be a good approximation to it, but its statistical properties will be more intimately involved in the inheritance of real stature than the properties of that variate itself. Since we are only concerned with variation we may take as a primary ingredient of the value of X, the mean value of x in the population, and adjust our positive and negative increments for each factor so that these balance each other when the whole population is considered. Since the increment for any one gene will appear p times to that for its alternative gene q times in the whole population, the two increments must be of opposite sign and in the ratio $q : (-p)$. Moreover, since their difference must be a, the actual values cannot but be qa and $(-pa)$ respectively.

The value of the *average excess a* of any gene substitution was obtained by comparing the average values of the measurement x in two moieties into which the population can be divided. It is evident that the values of a will only be properly determined if the same average difference is maintained in these moieties between the values of X, or in other words if in each such moiety the sum of the deviations, $x - X$, is zero. This supplies a criterion mathematically sufficient to determine the values of a, which represent in the population concerned the average effects of the gene substitutions. It follows that the sum for the whole population of the product $X(x - X)$ derived from each individual must be zero, for each entry qa or $(-pa)$ in the first term will in the total be multiplied by a zero, and this will be true of the items contributed by every factor severally. It follows from this that if X and x are now each measured from the mean of the population, the variance of X, which is the mean value of X^2, is equal to the mean value of Xx. Now the mean value of Xx will involve a for each Mendelian factor; for X will contain the item qa in the p individuals of one moiety and $(-pa)$ in the q individuals of the other, and since the average values of x in these two moieties differ by a and each individual contains two genes at each locus, the mean value of Xx must be the sum for all factors of the quantities $2pqa\alpha$. Thus the variance of X is shown to be

$$W = \Sigma(2pqa\alpha)$$

the summation being taken over all factors, and this quantity we may distinguish as the *genetic* variance in the chosen measurement x. That it is essentially positive, unless the effect of every gene severally is zero, is shown by its equality with the variance of X. An extension of this analysis, involving no difference of principle, leads to a similar expression for cases in which one or more factors have more than two different genes or allelomorphs present.

The appropriateness of the term genetic variance lies in the fact that the quantity X is determined solely by the genes present in the individual, and is built up of the average effects of these genes. It therefore represents the genetic potentiality of the individual concerned, in the aggregate of the mating possibilities actually open to him, in the sense that the progeny averages (of x, as well as of X) of two males mated with an identical series of representative females will differ by exactly half as much as the genetic potentialities of their sires differ. Relative genetic values may therefore be determined experimentally by the diallel method, in which each animal tested is mated to the same series of animals of the opposite sex, provided that a large number of offspring can be obtained from each such mating, and that the mates are representative of the actual population. Without obtaining individual values, the genetic variance of the population may be derived from the correlations between relatives, provided these correlations are accurately obtained.   For this purpose the square of the parental correlation divided by the grandparental correlation supplies a good estimate of the fraction, of the total observable variance of the measurement, which may be regarded as genetic variance.

It is clear that the actual measurements, x, obtained in individuals may differ from their genetic expectations by reason of fluctuations due to purely environmental circumstances. It should be noted that this is not the only cause of difference, for even if environmental fluctuations were entirely absent, and the actual measurements therefore determined exactly by the genetic composition, these measurements, which may be distinguished as *genotypic*, might still differ from the genetic values, X. A good example of this is afforded by dominance, for if dominance is complete the genotypic value of the heterozygote will be exactly the same as that of the corresponding dominant homozygote, and yet these genotypes differ by a gene substitution which may materially affect the genetic potentiality represented by X, and be reflected in the average measurement of the offspring. A similar cause of discrepancy occurs when gene substitutions in different factors are not exactly additive in their average effects. The genetic variance as here defined is only a portion of the variance determined genotypically, and this will differ from, and usually be somewhat less than, the total variance to be observed.

It is consequently not a superfluous refinement to define the purely genetic element in the variance as it exists objectively, as a statistical character of the population, different from the variance derived from the direct measurement of individuals.

Often more than two genes may alternatively occupy the same locus. These are termed multiple allelomorphs. In extending the notion of genetic excess to such cases, it is convenient to define the genetic excess associated with a single gene. Thus if we suppose that the genotypic value X has been ascertained for an entire natural population, the genetic composition of each individual of which is known, we may let $\overline{X}$ stand for the general mean, and x for the deviation of any genotypic value, so that

$$x = X - \overline{X}.$$

Choosing any particular factor, we may pick out all the individuals carrying any one gene, counting the homozygotes twice, and find the average value of x for this selected group of individuals.

Thus if out of a population of N individuals there are n_{11} homozygotes, and n_{lk} heterozygotes formed by combination with any other chosen allelomorph, the total of the values of x from the homozygotes may be represented by $S(x_{11})$, and that from any class of heterozygotes containing the chosen gene by $S(x_{lk})$. Then

$$\frac{2S(n_{11}) + \sum\limits_{k=2}^{s}{}'S(n_{lk})}{2n_{11} + \sum\limits_{k=2}^{s}{}'n_{lk}} = a_1$$

where a_1 may be spoken of as the average genotypic excess of the particular gene chosen. Σ is used for summation over allelomorphs of the same factor. If p_1 is the proportion of this kind of gene among all homologous kinds which might occupy the same locus, it is evident that

$$\sum\limits_{k=1}^{s}(p_k a_k) = 0.$$

We may now introduce a second quantity which has sometimes been confused with that defined above, and which may indeed have a nearly equal value, yet which must be distinguished from it in an accurate argument; namely the average effect produced in the population genetically constituted by the substitution of one gene for another. With multiple allelomorphism it is convenient to define this quantity also by the effect of substituting any chosen gene for a random selection of the genes homologous with it.   By whatever rules mating, and consequently the frequency of different combinations, may be governed, the substitution of a small proportion of genes of one kind for the others will produce a definite proportional effect upon the average measurement.   The amount of the difference produced, on the average, in the total measurement of the population, for each such gene substitution, may be termed the *average effect* of the gene substituted in contradistinction to the average excess defined above.   In human stature, for example, the correlation found between married persons is sufficient to ensure that each gene must be associated with other genes having a like effect to an extent sufficient to make the average excess associated with each gene exceed its average effect by about a quarter.

If α is the magnitude of the average effect of a given gene, its measurement by direct substitution implies the requirement that the values of α are the best additive system for predicting the genotypic value from the actual genes present in any individual.   Subject to the condition

$$\sum_{k=1}^{s}{}'(p\alpha) = 0$$

let ξ represent the value of the genotype as best predicted from the genes present, so that

$$\xi = \Sigma(\alpha)$$

where Σ stands for summation over all the genetic factors affecting stature.   Then if we make

$$S(x - \xi)^2$$

a minimum for variation of all quantities α, we find varying any one of them, α_1,

$$2S(x_{11} - \xi_{11}) + \sum_{k=2}^{s}{}' S(x_{ik} - \xi_{lk}) = \lambda p_1$$

where λ is some constant undetermined.   These equations are sufficient to determine all values of α.

Consider now the quantity

$$S\{\xi(x - \xi)\} = S(\xi x) - S(\xi^2) ;$$

if we substitute for the first ξ in the expression on the left its expression in terms of α, the coefficient of any particular value such as α_1 is

$$2S(x_{11} - \xi_{11}) + \sum_{k=2}^{s} S(x_{ik} - \xi_{1lk})$$

which we have shown to be equal to λp_1. Hence

$$S(\xi x) - S(\xi^2) = \sum_{k=1}^{s}{}'(p_k \alpha_k) = 0 .$$

It follows that the variance of the genetic value ξ is equal to the co-variance of the genetic and genetypic values ξ and x.

If now we substitute for ξ in the expression $S(\xi x)$ the coefficient of α_1 is

$$2S(x_{11}) + \sum_{k=2}^{s}{}'S(x_{lk}) = 2N p_1 a_1$$

by definition of a. The total of the contributions from any set of homologous genes is therefore

$$2N \sum_{k=1}^{s}(p_k a_k \alpha_k)$$

and for all factors affecting stature it is

$$2N \Sigma \Sigma'(pa\alpha) .$$

Dividing by N, the total number of individuals involved, it is now seen that the genetic variance of diploid individuals is given by

$$\Sigma \Sigma'(2pa\alpha) .$$

If only two allelomorphs are present, we have

$$p_1 + p_2 = 1$$
$$p_1 a_1 + p_2 a_2 = 0$$

whence, if $a_1 - a_2 = d$, then

$$a_1 = p_2 d \quad \text{and} \quad a_2 = - p_1 d .$$

Similarly, where $\alpha_1 = p_2 \delta$ and $\alpha_2 = - p_2 \delta$, then

$$\Sigma'(2pa\alpha) = 2p_1 a_1 \alpha_1 + 2p_2 a_2 \alpha_2$$
$$= 2p_1 p_2^2 \, d\delta + 2p_1^2 p_2 \, d\delta$$
$$= 2p_1 p_2 \, d\delta$$

in accordance with the expression obtained in the first edition of this book for the case of only two allelomorphs. In the formula there given the factor 2 for diploids is omitted, through treating the population as one of the $2N$ loci instead of N individuals, as seems to be in every way preferable.

Natural Selection

Any group of individuals selected as bearers of a particular gene, and consequently the genes themselves, will have rates of increase which may differ from the average. The excess over the average of any such selected group will be represented by a, and similarly the average effect upon m of introducing the gene in question will be represented by α. Since m measures fitness by the objective fact of representation in future generations, the quantity

$$\Sigma'(2pa\alpha)$$

will represent the contribution of each factor to the genetic variance in fitness. The total genetic variance in fitness being the sum of these contributions, which is necessarily positive, or, in the limiting case, zero. Moreover, any increase dp in the frequency of the chosen gene will be accompanied by an increase $2\alpha\,dp$ in the average fitness of the species, where α may, of course, be negative. But the definition of a requires that

$$\frac{d}{dt}\log p = a$$

or
$$dp = (pa)dt$$

hence
$$(2\alpha)dp = (2pa\alpha)dt$$

which must represent the rate of increase of the average fitness due to the change in progress in frequency of this one gene. Summing for all allelomorphic genes, we have

$$dt\Sigma'(2pa\alpha)$$

and taking all factors into consideration, the total increase in fitness is

$$\Sigma\alpha\,dp = dt\Sigma\Sigma'(2pa\alpha) = W\,dt\,.$$

If therefore the time element dt is positive, the total change of fitness Wdt is also positive, and indeed the rate of increase in fitness due to all changes in gene ratio is exactly equal to the genetic variance of fitness W which the population exhibits. We may consequently state the fundamental theorem of Natural Selection in the form:

The rate of increase in fitness of any organism at any time is equal to its genetic variance in fitness at that time.

The rigour of the demonstration requires that the terms employed should be used strictly as defined; the ease of its interpretation may be increased by appropriate conventions of measurement. For example, the frequencies p should strictly be evaluated at any instant by the enumeration, not necessarily of the census population, but of all individuals having reproductive value, weighted according to the reproductive value of each.

Since the theorem is exact only for idealized populations, in which fortuitous fluctuations in genetic composition have been excluded, it is important to obtain an estimate of the magnitude of the effect of these fluctuations, or in other words to obtain a standard error appropriate to the calculated, or expected, rate of increase in fitness. It will be sufficient for this purpose to consider the special case of a population mating and reproducing at random. It is easy to see that if such chance fluctuations cause a difference δp between the actual value of p obtained in any generation and that expected, the variance of δp will be

$$\frac{pq}{2n},$$

where n represents the number breeding in each generation, and $2n$ therefore is the number of genes in the n individuals which live to replace them. The variance of the increase in fitness, $\Sigma 2\alpha dp$, due to this cause, will therefore be

$$\frac{1}{2n}(2pq\alpha^2).$$

Now, with random mating, the chance fluctuation in the different gene ratios will be independent, and the values of a and α are no longer distinct, it follows that, on this condition, the rate of increase of fitness, when measured over one generation, will have a standard error due to random survival equal to

$$\frac{1}{T}\sqrt{\frac{W}{2n}}$$

where T is the time of a generation. It will usually be convenient for each organism to measure time in generations, and if this is done it will be apparent from the large factor $2n$ in the denominator, that the random fluctuations in W, even measured over only a single generation, may be expected to be very small compared to the average rate of progress. The regularity of the latter is in fact guaranteed by the same circumstance which makes a statistical assemblage of particles, such as a bubble of gas obey, without appreciable deviation, the laws of gases. A visible bubble will indeed contain several billions of molecules, and this would be a comparatively large number for an organic population, but the principle ensuring regularity is the same. Interpreted exactly, the formula shows that it is only when the rate of progress, W, when time is measured in generations, is itself so small as to be comparable to $1/n$, that the rate of progress achieved in successive generations is made to be irregular. Even if an equipoise of this order of exactitude, between the rates of death and reproduction of different genotypes, were established, it would be only the rate of progress for spans of a single generation that would be shown to be irregular, and the deviations from regularity over a span of 10,000 generations would be just a hundredfold less.

It will be noticed that the fundamental theorem proved above bears some remarkable resemblances to the second law of thermodynamics. Both are properties of populations, or aggregates, true irrespective of the nature of the units which compose them; both are statistical laws; each requires the constant increase of a measurable quantity, in the one case the entropy of a physical system and in the other the fitness, measured by m, of a biological population. As in the physical world we can conceive of theoretical systems in which dissipative forces are wholly absent, and in which the entropy consequently remains constant, so we can conceive, though we need not expect to find, biological populations in which the genetic variance is absolutely zero, and in which fitness does not increase. Professor Eddington has recently remarked that 'The law that entropy always increases—the second law of thermodynamics—holds, I think, the supreme position among the laws of nature'. It is not a little instructive that so similar a law should hold the supreme position among the biological sciences. While it is possible that both may ultimately be absorbed by some more general principle, for the present we should note that the laws as they stand present profound differences— (1) The systems considered in thermodynamics are permanent; species on the contrary are liable to extinction, although biological improvement must be expected to occur up to the end of their existence. (2) Fitness, although measured by a uniform method, is qualitatively different for every different organism, whereas entropy, like temperature, is taken to have the same meaning for all physical systems. (3) Fitness may be increased or decreased by changes in the environment, without reacting quantitatively upon that environment. (4) Entropy changes are exceptional in the physical world in being irreversible, while irreversible evolutionary changes form no exception among biological phenomena. Finally, (5) entropy changes lead to a progressive disorganization of the physical world, at least from the human standpoint of the utilization of energy, while evolutionary changes are generally recognized as producing progressively higher organization in the organic world.

The statement of the principle of Natural Selection in the form of a theorem determining the rate of progress of a species in fitness to survive (this term being used for a well-defined statistical attribute of the population), together with the relation between this rate of progress and its standard error, puts us in a position to judge of the validity of the objection which has been made, that the principle of Natural Selection depends on a succession of favourable chances. The objection is more in the nature of an innuendo than of a criticism, for it depends for its force upon the ambiguity of the word chance, in its popular uses. The income derived from a Casino by its proprietor may, in one sense, be said to depend upon a succession of favourable chances, although the phrase contains a suggestion of improbability more appropriate to the hopes of the patrons of his establishment. It is easy without any very profound logical analysis to perceive the difference between a succession of favourable deviations from the laws of chance, and on the other hand, the continuous and cumulative action of these laws. It is on the latter that the principle of Natural Selection relies.

In addition to the genetic variance of any measurable character there exists, as has been seen, a second element comprised in the total genotypic variance, due to the heterozygote being in general not equal to the mean of the two corresponding homozygotes. This component, ascribable to dominance, is also in a sense capable of exerting evolutionary effects, not through any direct effect on the gene ratios, but through its possible influence on the breeding system. For if, in general, heterozygotes were favoured as compared with homozygotes, it is evident that the offspring of outcrosses would be at an advantage compared with those of matings between relatives, or of self-fertilization, and any heritable tendencies favouring such matings might come to be eliminated, with consequent increase in the proportion of heterozygotes.

This indirect and conditional factor in selection seems to have been able to produce effects of considerable importance, such as the separation of the sexes, self-sterility in many plants, and flowers made attractive by colour, scent and nectar.   A first step to the understanding of these effects of dominance has been made in Chapter III, but the author would emphasize that in his opinion no satisfactory selective model has been set up competent even to derive a distylic species like the primrose from a monostylic species of the same genus.   Possibly, therefore, the course of evolutionary change has been complex and circuitous.

Such effects ascribable to the dominance component of the genotypic variation are not in reality additional to the evolutionary changes accounted for by the fundamental theorem; for in that theorem they are credited to the gene-substitutions needed, for example, to develop bigger or brighter flowers; although the selective advantage conferred by these may be wholly due to dominance deviations in fitness recognizable in numerous other factors.

The nature of adaptation

In order to consider in outline the consequences to the organic world of the progressive increase of fitness of each species of organism, it is necessary to consider the abstract nature of the relationship which we term 'adaptation'.   This is the more necessary since any *simple* example of adaptation, such as the lengthened neck and legs of the giraffe as an adaptation to browsing on high levels of foliage, or the conformity in average tint of an animal to its natural background, lose, by the very simplicity of statement, a great part of the meaning which the word really conveys.   For the more complex the adaptation, the more numerous the different features of conformity, the more essentially adaptive the situation is recognized to be.   An organism is regarded as adapted to a particular situation, or to the totality of situations which constitute its environment, only in so far as we can imagine an assemblage of slightly different situations, or environ-ments, to which the animal would on the whole be less well adapted; and equally only in so far as we can imagine an assemblage of slightly different organic forms, which would be less well adapted to that environment.   This I take to be the meaning which the word is intended to convey, apart altogether from the question whether organisms really are adapted to their environments, or whether the structures and instincts to which the term has been applied are rightly so described.

The statistical requirements of the situation, in which one thing is made to conform to another in a large number of different respects, may be illustrated geometrically. The degree of conformity may be represented by the closeness with which a point A approaches a fixed point O. In space of three dimensions we can only represent conformity in three different respects, but even with only these the general character of the situation may be represented. The possible positions representing adaptations superior to that represented by A will be enclosed by a sphere passing through A and centred at O. If A is shifted through a fixed distance, r, in any direction its translation will improve the adaptation if it is carried to a point within this sphere, but will impair it if the new position is outside. If r is very small it may be perceived that the chances of these two events are approximately equal, and the chance of an improvement tends to the limit $\frac{1}{2}$ as r tends to zero; but if r is as great as the diameter of the sphere or greater, there is no longer any chance whatever of improvement, for all points within the sphere are less than this distance from A. For any value of r between these limits the actual probability of improvement is

$$\frac{1}{2}\left(1 - \frac{r}{d}\right),$$

where d is the diameter of the sphere.

The chance of improvement thus decreases steadily from its limiting value $\frac{1}{2}$ when r is zero, to zero when r equals d. Since A in our representation may signify either the organism or its environment, we should conclude that a change on either side has, when this change is extremely minute, an almost equal chance of effecting improvement or the reverse; while for greater changes the chance of improvement diminishes progressively, becoming zero, or at least negligible, for changes of a sufficiently pronounced character.

The representation in three dimensions is evidently inadequate; for even a single organ, in cases in which we know enough to appreciate the relation between structure and function, as is, broadly speaking, the case with the eye in vertebrates, often shows this conformity in many more than three respects. It is of interest therefore, that if in our geometrical problem the number of dimensions be increased, the form of the relationship between the magnitude of the change r and the probability of improvement, tends to a limit which is represented in Fig. 3. The primary facts of the three dimensional problem are conserved in that the chance of improvement, for very small displacements tends to the limiting value $\frac{1}{2}$, while it falls off rapidly for increasing displacements, attaining exceedingly small values, however, when the number of dimensions is large, even while r is still small compared to d.

For any degree of adaptation there will be a standard magnitude of change, represented by $d/\sqrt{n}$, and the probability of improvement will be determined by the ratio which the particular change considered bears to this standard magnitude. The higher the adaptation the smaller will this standard be, and consequently the smaller the probability that a change of given magnitude shall effect an improvement. The situation may be expressed otherwise by supposing changes of a given magnitude to occur at random in all directions, and comparing the rates of evolutionary progress caused by two opposite selective agencies, one of which picks out and accumulates all changes which increase the adaptation, and another which similarly picks out and accumulates all which diminish it. For changes very small compared to the standard, these two agencies will be equally effective, but, even for changes of only one-tenth of the standard, the destructive selection is already 28 per cent. more effective than the selection favouring

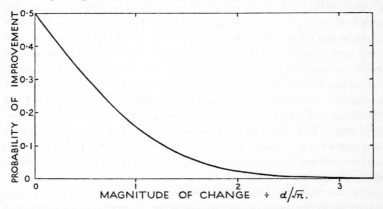

Fig. 3. The relation between the magnitude of an undirected change and the probability of improving adaptation, where the number of dimensions (n) is large

$$p = \frac{1}{\sqrt{2\pi}} \int_{x}^{\infty} e^{-\frac{1}{2}t^2} dt, \; x = r\sqrt{n}/d.$$

adaptation. At one half the standard it is over three and a half times as powerful, at the standard value itself, at which the probability of improvement is still, as the diagram shows, nearly one in six, the selection destroying adaptation is thirteen times as effective as that building it up, and at twice and three times the standard value the ratio has risen to the values 236 and 7,852 respectively.

The conformity of these statistical requirements with common experience will be perceived by comparison with the mechanical adaptation of an instrument, such as the microscope, when adjusted for distinct vision. If we imagine a derangement of the system by moving a little each of the lenses, either longitudinally or transversely, or by twisting through an angle, by altering the refractive index and transparency of the different components, or the curvature, or the polish of the interfaces, it is sufficiently obvious that any large derangement will have a very small probability of improving the adjustment, while in the case of alterations much less than the smallest of those intentionally effected by the maker or the operator, the chance of improvement should be almost exactly one half.

Deterioration of the environment

If therefore an organism be really in any high degree adapted to the place it fills in its environment, this adaptation will be constantly menaced by any undirected agencies liable to cause changes to either party in the adaptation. The case of large mutations to the organism may first be considered, since their consequences in this connexion are of an extremely simple character. A considerable number of such mutations have now been observed, and these are, I believe, without exception, either definitely pathological (most often lethal) in their effects, or with high probability to be regarded as deleterious in the wild state. This is merely what would be expected on the view, which was regarded as obvious by the older naturalists, and I believe by all who have studied wild animals, that organisms in general are, in fact, marvellously and intricately adapted, both in their internal mechanisms, and in their relations to external nature. Such large mutations occurring in the natural state would be unfavourable to survival, and as soon as the numbers affected attain a certain small proportion in the whole population, an equilibrium must be established in which the rate of elimination is equal to the rate of mutation. To put the matter in another way we may say that each mutation of this kind is allowed to contribute exactly so much to the genetic variance of fitness in the species as will provide a rate of improvement equivalent to the rate of deterioration caused by the continual occurrence of the mutation.

As to the physical environment, geological and climatological changes must always be slowly in progress, and these, though possibly beneficial to some few organisms, must as they continue become harmful to the greater number, for the same reasons as mutations in the organism itself will generally be harmful. For the majority of organisms, therefore, the physical environment may be regarded as constantly deteriorating, whether the climate, for example, is becoming warmer or cooler, moister or drier, and this will tend, in the majority of species, constantly to lower the average value of m, the Malthusian parameter of the population increase. Probably more important than the changes in climate will be the evolutionary changes in progress in associated organisms. As each organism increases in fitness, so will its enemies and competitors increase in fitness; and this will have the same effect, perhaps in a much more important degree, in impairing the environment, from the point of view of each organism concerned. Against the action of Natural Selection in constantly increasing the fitness of every organism, at a rate equal to the genetic variance in fitness which that population maintains, is to be set off the very considerable item of the deterioration of its inorganic and organic environment. This at least is the conclusion which follows from the view that organisms are very highly adapted. Alternatively, we may infer that the organic world in general must tend to acquire just that level of adaptation at which the deterioration of the environment is in some species greater, though in some less, than the rate of improvement by Natural Selection, so as to maintain the general level of adaptation nearly constant.

Changes in population

An increase in numbers of any organism will impair its environment in a manner analogous to, and more surely than, an increase in the numbers or efficiency of its competitors. It is a patent oversimplification to assert that the environment determines the numbers of each sort of organism which it will support. The numbers must indeed be determined by the elastic quality of the resistance offered to increase in numbers, so that life is made somewhat harder to each individual when the population is larger, and easier when the population is smaller. The balance left over when from the rate of increase in the mean value of

m produced by Natural Selection, is deducted the rate of decrease due to deterioration in environment, results not in an increase in the average value of m, for this average value cannot greatly exceed zero, but principally in a steady increase in population.

The situation is represented by the differential equation

$$\frac{dM}{dt} + \frac{M}{C} = W - D$$

in which M is the mean of the Malthusian parameter, C is a constant expressing the relation between fitness and population increase, and defined as the increase in the natural logarithm of the population, supposed stationary at each stage, produced by unit increase in the value of M, W is the rate of actual increase in fitness determined by natural selection, and D is the rate of loss due to the deterioration of the environment. If C, W and D are constant the equation has the solution

$$M = C(W - D) + Ae^{-t/C}$$

in which A is an arbitrary constant, determined by the initial conditions. C has the physical dimensions of time, and may therefore be reckoned in years or generations, and the equation shows that if C, W, and D remain constant for any length of time much greater than C, the value of M will approach to the constant value given by

$$M = C(W - D).$$

In this steady state the whole of the organism's advantage or disadvantage will be compensated by change in population, and not at all by change in the value of M.

A word should perhaps be said as to the form of statement of selection theory which ascribes the 'struggle for existence' to the excessive production of offspring, supposedly to be observed throughout organic nature. If the numbers of a species are adjusted to that level at which each adult produces on the average just two offspring which attain the adult state, then, if there is any mortality whatever in the previous life stages, either through inorganic causes, or by reason of predators and parasites, it necessarily follows that young must be produced in excess of the parental numbers. If the mortality is high, then the ratio of this excess will be large. Having realized this situation, if we now imagine an ideal world in which all these offspring attain maturity and breed, it is obvious that in such a world the numbers of the species considered will increase without limit. It is usually added, though this is logically irrelevant, that the increase will be in geometrical progression. We may in this sense speak of the production of offspring as 'excessive', and the geometrical rate of increase with its impressive picture of over-population, has been widely represented as a logical basis of the argument for natural selection. However, it should be remembered that the production of offspring is only excessive in relation to an imaginary world, and the 'high geometrical rate of increase' is only attained by abolishing a real death rate, while retaining a real rate of reproduction. There is something like a relic of creationist philosophy in arguing from the observation, let us say, that a cod spawns a million eggs, that *therefore* its offspring are subject to Natural Selection; and it has the disadvantage of excluding fecundity from the class of characteristics of which we may attempt to appreciate the aptitude. It would be instructive to know not only by what physiological mechanism a just apportionment is made between the nutriment devoted to the gonads and that devoted to the rest of the parental organism, but also what circumstances in the life-history and environment would render profitable the diversion of a greater or lesser share of the available resources towards reproduction. The historical fact that both Darwin and Wallace were led through reading Malthus's essay on population to appreciate the efficacy of selection, though extremely instructive as to the philosophy of their age, should no longer constrain us to confuse the consequences of that principle with its foundations.

It will have been apparent in the earlier sections of this chapter that the actuarial information necessary for the calculation of the genetic changes actually in progress in a population of organisms, will always be lacking; if only because the number of different genotypes for each of which the Malthusian parameter is required will often, perhaps always, exceed the number of organisms in the population, in addition to the fact that this parameter is very imperfectly known even in human population aggregates, for which vital statistics are in some degree available. If, however, we are content to consider not in full detail exactly what changes are in progress, but quite broadly to what extent an organism is holding its own in the economy of nature, it is only necessary to determine the numerical values of the four quantities W, D, C, and M, which enter into the equation of population growth. Our ignorance as to these is, of course, profound, but, regarding the problem in this limited aspect, it is by no means obvious, with respect to organisms of sufficient importance to deserve detailed study, that it could not largely be removed by systematic and well-directed observations. The quantity C, for example, which is a period of time, measuring the facility with which, with increased fitness, the population is allowed to increase, must be intimately related to the course of population increase or decrease, with which the numbers of an organism exposed to new influences, approach an equilibrium value, which over short periods may be regarded as stationary. An organism introduced into a new environment, to which it is well suited, will increase in numbers rapidly for a comparatively few years, and somewhat rapidly attain its equilibrium density. The same must be true of the decrease of a population exposed by man to new causes of destruction. In these cases it is probable that the process of attaining equilibrium is sufficiently rapid for the changes due to organic evolution, and the natural deterioration of the environment, to be neglected, and further changes in the extent of human intervention could, for experimental purposes, be suspended locally. In such cases, at their simplest, the course of population change would be represented by the equation $M = Ae^{-t/C}$, or, since M is the relative or logarithmic growth rate of the population, by $\log N = \log N_0 - ACe^{-t/C}$, where N is the size or density of the population, and N_0 the steady value to which it is tending. Observations of N will then determine, at least approximately, the value of the time constant C. It should be noticed that for such comparatively large changes of population density as could be measured with sufficient precision, important changes will often take place in the numbers of associated organisms. The simple relation obtained above will only be satisfactory if these associated changes take place rapidly in comparison to the change we are studying. Otherwise it would be necessary to take account by direct observation of the changes in numbers of at least the more important of the associated organisms, and so to determine the constants of the more complex system of differential equations by which their interactions may be represented.

With respect to the other constants, the practical difficulties appear to be greater, though, seeing how little attention in general has been paid to the quantitative study of organisms in their natural habitats, it would be rash to assume that their determination is beyond human endeavour. It is certainly possible by continuously sustained research to ascertain the densities of population of animals and plants, and to study their variations in relation to the climatic and other environmental factors of their habitats. Knowledge of this kind serves to indicate to what extent physical changes now in progress can be improving or impairing the environment. Given the time constant C, these effects may be translated directly into terms of fitness. In certain cases, such as the slow changes in composition of plant associations, the value of M might be directly determined, and in conjunction with more or less trustworthy determinations of C and D, this would lead to a more or less exact estimate of the evolutionary factor W. The direct determination of the latter quantity would seem to require a complete genealogy of the species for several generations, and this can only be possible in Man. Moreover, owing to the rapid changes which man is making in his environment, it may be foreseen that human genealogies on a national or international scale, such as are certainly now possible, while throwing an immense amount of light on the current conditions of human reproduction and survival, will offer special difficulties in the determination and interpretation of the evolutionary value W.

'The benefit of the species'

It will be observed that the principle of Natural Selection, in the form in which it has been stated in this chapter, refers only to the variation among individuals (or co-operative communities), and to the progressive modification of structure or function only in so far as variations in these are of advantage to the individual, in respect to his chance of death or reproduction. It thus affords a rational explanation of structures, reactions and instincts which can be recognized as profitable to their individual possessors. It affords no corresponding explanation for any properties of animals or plants which, without being individually advantageous, are supposed to be of service to the species to which they belong.

This distinction was unknown to the early speculations to which the perfection of adaptive contrivances naturally gave rise. For the interpretation that these were due to the particular intention of the Creator would be equally appropriate whether the profit of the individual or of the species were the objective in view. The phrases and arguments of this pre-Darwinian viewpoint have, however, long outlived the philosophy to which they belong. It would be easy to find among modern writers many parallels to the thought expressed in the following quotation:—

'Of what advantage could it be to any species for the males to struggle for the females and for the females to struggle for the males?'

This sort of question might appropriately be put to an opponent who claimed that the instincts of animals were in each case due to the direct contrivance of the Creator. As a means of progressive change, on the contrary, Natural Selection can only explain these instincts in so far as they are individually beneficial, and leaves entirely open the question as to whether in the aggregate they are a benefit or an injury to the species.

There would, however, be some warrant on historical grounds for saying that the term Natural Selection should include not only the selective survival of individuals of the same species, but of mutually competing species of the same genus or family. The relative unimportance of this as an evolutionary factor would seem to follow decisively from the small *number* of closely related species which in fact do come into competition, as compared to the number of individuals in the same species; and from the vastly greater *duration* of the species compared to the individual. Any characters ascribed to interspecific selection should of course characterize, not species, but whole genera or families, and it may be doubted if it would be possible to point to any such character, with the possible exception, as suggested in Chapter VI, of sexuality itself, which could be interpreted as evolved for the specific rather than for the individual advantage.

Summary

The vital statistics of an organism in relation to its environment provide a means of determining a measure of the relative growth-rate of the population, which may be termed the Malthusian parameter of population increase, and provide also a measure of the reproductive values of individuals at all ages or stages of their life-history. The Malthusian parameter will in general be different for each different genotype, and will measure the fitness to survive of each.

The variation in a population of any individual measurement is specified quantitatively by its variance, and of this, taking account of the genetic composition of all possible individuals, a definite amount may be recognized as genetic variance.

The rate of increase of fitness of any species is equal to the genetic variance in fitness, and the standard error of this rate of progress even over a single generation, will (unless the latter is so exceedingly minute as to be comparable, when time is measured in generations, to the reciprocal of the number of organisms in the population) be small compared to the rate of progress.

Adaptation, in the sense of conformity in many particulars between two complex entities, may be shown, by making use of the geometrical properties of space of many dimensions, to imply a statistical situation in which the probability, of a change of given magnitude effecting an improvement, decreases from its limiting value of one half, as the magnitude of the change is increased. The intensity of adaptation is inversely proportional to a standard magnitude of change for which this probability is constant. Thus the larger the change, or the more intense the adaptation, the smaller will be the chance of improvement.

Against the rate of progress in fitness must be set off, if the organism is, properly speaking, highly adapted to its place in nature, deterioration due to undirected changes either in the organism, or in its environment. The former, typified by the pathological mutations observed by geneticists, annul their influence by calling into existence an equivalent amount of genetic variance. The latter, which are due to geological and climatological changes on the one hand, and to changes in the organic environment, including the improvement of enemies and competitors, on the other, may be in effect either greater or less than the improvement due to Natural Selection.

Any net advantage gained by an organism will be conserved in the form of an increase in population, rather than in an increase in the average Malthusian parameter, which is kept by this adjustment always near to zero.

Although it appears impossible to conceive that the detailed action of Natural Selection could ever be brought completely within human knowledge, direct observational methods may yet determine the numerical values which condition the survival and progress of particular species.

III

THE EVOLUTION OF DOMINANCE

The dominance of wild genes. Modification of the effects of Mendelian factors. Modification of the heterozygote. Special applications of the theory. The process of modification. Inferences from the theory of the evolution of dominance. Summary.

The very object of hypothesis is to inquire whether a real cause has not had a wider operation than there is any direct evidence for. ROBERTSON SMITH.

The dominance of wild genes

IT has been seen in Chapter I that it is scarcely possible, in the light of the particulate nature of inheritance, to ascribe to mutations any importance in determining the direction of evolutionary change; their importance in evolution lies in playing the very different role of maintaining the stock of genetic variance at a certain level, which level in its turn is a factor in determining the speed, though not the direction, of evolutionary progress. Before attempting to consider in detail the relations between the amount of the stock of genetic variability in a species, the rates of mutation, and the size of the population, as will be done in Chapter IV, it is necessary to examine, as far as the present state of the evidence allows, into the character of the genetic changes known as mutations.

The principle of this chapter was novel enough in 1930 to require a good deal of explanation or apology. In the light of modern knowledge, however, it is seen to be one only, but one of the most interesting, of the aspects of a much wider principle, namely that the effects by which any gene-substitution is recognized depend on the results of interactions with, possibly, all other ingredients of the germ plasm, and so may be altered or abolished by changes in these latter. Although this wider principle is now recognized, yet it was the need to give an evolutionary interpretation of the genetic facts regarding dominance that was the first occasion of its recognition. If after twenty-five years this chapter has required rewriting to a greater extent than others, this is chiefly due to the progress since made in this direction, so that the tentative and apologetic approach which seemed appropriate in 1930 would now be quite misleading.

The term mutation has been applied to a number of different kinds of intracellular events, having in common the production of heritable novelties. Cases are known of the doubling of the entire chromosome outfit, the doubling of single chromosomes, and of parts of chromosomes; in other cases a part of a chromosome appears to be translocated from its habitual site and attached to some other chromosome; and these are all mutations in the wide and primitive meaning of the term. Nevertheless, the evolutionary possibilities of these kinds of change are evidently extremely limited compared to those of the type of change to which the term gene-mutation is applied. This consists in a change in a single hereditary particle, or gene, into a gene of a new type, occupying the same locus in the germinal structure. The grosser forms of mutation may indeed play a special evolutionary role in supplying a mechanism of reproductive incompatibility, which may be of importance when physiological isolation is in question, but only in very special cases could they contribute appreciably to the genetic diversity of an interbreeding population.

With respect to any pair of alternative or allelomorphic genes, the one may be distinguished from the other in four different respects, which, in order to examine their relationships, it is important to keep conceptually distinct. We may distinguish (a) the rarer from the more common, (b) the less advantageous from the more advantageous, (c) the mutant gene from the relatively primitive gene from which it arose, and finally (d) the recessive gene from the dominant. It is only when these four means of contrast are kept distinct that we can appreciate the associations between them which arise from different causes.

In connexion with the nature of adaptation it has been seen that mutant genes will more often than not be disadvantageous, and that this will be most conspicuously the case with the factors having a large effect, and which consequently are more easily detected and studied. Distinctions (b) and (c), the quality of being less advantageous, and the quality of being the mutant as opposed to the parent gene, are thus very generally associated. With respect to rarity, if we consider a sequence of events in which a new gene, at first very rare, increases in frequency until it finally supplants its predecessor throughout the entire species, with but few exceptions, then in the first half of the process the commoner and in the second half the rarer will have been at a selective disadvantage. There is thus no general or universal association between the distinctions (a) and (b). Nevertheless, the great majority of mutations achieve no such success. A common genetic situation is that of an approximate equilibrium between mutations tending to increase the frequency of a disadvantageous gene, and counter-selection tending to eliminate it. In such circumstances even very gentle counterselection will suffice to make sure that the disadvantageous mutation is ordinarily found to be much the rarer.

In this situation the quality of dominance has an important influence. If, for example, the elimination of one recessive, homozygous for a mutant gene, in each million of each generation were sufficient to counterbalance the mutation rate, then a dominant gene with the same mutation rate would only have to maintain two heterozygotes in a million to encounter equivalent counterselection. But, to maintain a frequency of one in a million recessives in a cross-breeding population, the gene frequency would be one in a thousand, or, to express the matter otherwise, the typical sample of a million organisms would contain not only one homozygote, but in addition about two thousand heterozygotes carrying the mutant gene. The frequency, in the neighbourhood of equilibrium, of the recessive mutant gene would be a thousand times greater than that of a dominant gene having equivalent effect. The fact therefore that the rare recessives exposed by inbreeding are predominantly defects does not prove that mutant defects are generally recessive. This important point is, however, clearly demonstrated by other genetic observations.

Among species of plants propagated, like sweet peas, in distinct varieties, genetical analysis shows that the genes which on morphological grounds must be regarded as mutants, are, in an immense preponderance, recessives. In sweet peas complete recessiveness seems to be the invariable rule, as judged from the fifteen or twenty factors so far successfully elucidated. In the majority of such cases the occurrence of the mutation has not itself been observed, but the mutant gene is recognized as such by producing effects unknown in the older varieties, or wild prototypes. This case, with many others like it, constitutes very substantial and extensive evidence of the tendency of mutant genes to be recessive, for dominant mutants would be as eagerly seized upon and perpetuated as novelties, and would be more quickly detected than are recessives. The greater ease of detection is especially emphasized in the case of Man, where the recognition of a rare defect as due to a single Mendelian factor depends upon genealogical evidence; for the simplest pedigree, such as is almost always available, will reveal the character of a dominant defect, while the collation of the statistical evidence of extensive pedigree collections is usually necessary to demonstrate the Mendelian character of a simple recessive. For example, the Mendelian character even of albinism in Man has been disputed. It is a consequence of this difficulty that more dominant defects are known in Man than recessives, although there can be no doubt that the great bulk of human defects, physical and mental, are, as in other animals, recessive. This view is confirmed by the fact that sex-linked mutants, which *a priori* in an organism with 23 pairs of chromosomes should be a small minority, are prominent in the list of human defects; here the recessives are nearly as easily detected as dominants, and almost all known cases are recessive. In the case of so-called dominant defects in Man we only know that the heterozygote differs from the normal; we cannot ordinarily know what is the appearance of the homozygote mutant, or even if it is viable.

The imposing body of genetical researches devoted to the fruit-fly, *Drosophila*, in this as in other genetical questions has supplied the most decisive evidence. Something like 500 mutants have been actually observed in cultures of this fly, and, setting aside the large class of lethals, and all those which produce no visible effect, there remain 221 cases in which we can classify the mutation as Recessive, Intermediate, or, as it is usually called, Incompletely Dominant, or finally completely Dominant. The table shows the distribution into these three classes, respectively for the autosomal and the sex-linked mutations.

TABLE 1.

	Recessive.	Intermediate.	Completely Dominant.	Total.
Autosomal .	130	9	0	139
Sex-linked .	78	4	0	82

The classification has been compiled from the fine article on *Drosophila* in *Bibliographia Genetica*. In several individual instances my classification may be mistaken, either because the observations are affected by some uncertainty, or because they are not in every case explicitly stated. No one will doubt the extreme thoroughness with which the genetics of this fly has been investigated; it is, however, perhaps worth mentioning that whereas recessives are so common that the loss of one without investigation would seem no tragedy, mutations which can be 'used as dominant', that is the Intermediates, are of the greatest service to further research, and are much valued. They are, moreover, exposed to detection immediately upon their occurrence, whereas recessives are only noticed when they appear as homozygotes. For both reasons the proportion of recessives, high as it is, is likely to be an underestimate of the actual frequency of occurrence.

Lethal factors, which have been excluded from the enumeration set out above, provide independent though slightly equivocal confirmation of the same conclusion. The majority of so-called dominants are lethal in the homozygous condition, and must for this reason be properly classed as Intermediate. It is more remarkable that recessive lethals, which can produce no visible changes, were soon discovered by their effect in disturbing the frequency ratios of other factors, and especially of sex. It has since been demonstrated that both in normal conditions, and when mutations are artificially stimulated by X-rays, the recessive lethals are by far the most frequent class of mutation. Something like two per cent. of untreated fruit-flies must be mutants for some recessive lethal, and the frequency of mutations of this class must be quite tenfold that of all visible mutations. Whereas, among the latter, one in seventeen has been classed as intermediate in respect of dominance, the proportion must be even lower among lethals, unless indeed some of the obscure, though probably large, class of mutants which are lethal when heterozygous, be counted as dominant.

The pronounced tendency of the mutant gene to be recessive, to the gene of wild type from which it arises, calls for explanation, and there is fortunately an important group of observations available, to show that in this connexion we should stress the prevalence in the wild state of the dominant gene, rather than its relation of predecessor to the mutant which arises from it. Numerous cases are now known in which several different mutations have occurred to the same gene, and each of the mutant types can replace each other, and the wild type, in the same locus. In rodents, for example, several members of the albino series of genes have been found, ranging in effect from a slight dilution of pigmentation to its complete suppression. Using a set of five such alternative genes in the cavy or guinea pig, Sewall Wright has formed all of the fifteen possible combinations, five homozygous and ten heterozygous, which a set of five allelomorphs make possible. These he has examined in sufficient numbers to determine the average, and normal variation in depth of pigment both of the areas which range from black through sepia to white, and those which range from red through yellow. The four heterozygous forms containing the wild type gene are indistinguishable in depth of pigment from the homozygous wild type, from which all the other four homozygotes differ considerably. The remaining six heterozygous forms, which contain no gene of the wild type, all are clearly intermediate in both colours between the two homozygotes for the genes which they contain. This case, remarkable for the thoroughness with which it has been examined, is by no means exceptional. A number of similar series have been found in *Drosophila*, and the rule that the wild type gene dominates all others, but that these others show no mutual dominance, is stated as general by Morgan, Bridges, and Sturtevant.

The exceptional position in respect to dominance of the genes of the wild type among their allelomorphs is not owing to their being the originals from which the others arose by mutation, for one mutant allelomorph has been observed to arise from another, and mutant genes to mutate back to the wild type. We are driven therefore to see in dominance a characteristic proper, not to the predecessor as opposed to the successor in a series of mutational changes, but to the prevalent wild type as opposed to its unsuccessful competitors. Moreover, unless we are to abandon altogether the evolutionary conception of the modification of species by the occasional substitution of one gene for the predecessor from which it arose, the existence of the rule which gives genetical dominance to genes of the prevalent wild type requires that the successful new gene should in some way *become* dominant to its competitors, and if back mutations occur, to its predecessor also. The means by which this can occur are of special interest in the theory of Natural Selection, for they reveal an effect of selection which has nothing to do with its well-understood action in fitting a species to its place in nature. As has been indicated in Chapter II, it is scarcely possible to imagine a problem more intricate, or requiring so inconceivably detailed a knowledge of the bionomic situation, as that of tracing the net gain in fitness of any particular genetic change. Our knowledge in this respect, while sufficient to enable us to appreciate the adaptive significance of the differences in organization which distinguish whole orders or families, is almost always inadequate to put a similar interpretation on specific differences, and still more on intraspecific variation. This circumstance, which has been felt as a difficulty to the theory of Natural Selection by writers such as Bateson (1894) and Robson (1927), while admitting of notable exceptions, such as external colour, and especially the mimetic patterns of butterflies, does yet give an added interest to a case in which our quantitative information, while far from exact, is yet substantial and approximate.

Modification of the effects of Mendelian factors

The fashion of speaking of a given factor, or gene substitution, as causing a given somatic change, which was prevalent among the earlier geneticists, has largely given way to a realization that the change, although genetically determined, may be influenced or governed either by the environment in which the substitution is examined, or by the other elements in the genetic composition. Cases were fairly early noticed in which a factor, B, produced an effect when a second factor, A, was represented by its recessive gene, but not when the dominant gene was present. Factor A was then said to be epistatic to factor B, or more recently b would be said to be a specific modifier of a, where the small letters are taken to represent the recessive genes. There are other cases in which neither A nor B produce any effect when the other is recessive, in which cases we speak of the two factors as complementary; again neither may produce any effect if the other is dominant, when we speak of the two factors as duplicate. These are evidently only particular examples of the more general fact that the visible effect of a gene substitution depends both on the gene substitution itself and on the genetic complex, or organism, in which this gene substitution is made. We may perhaps find a form of words which reduces to a minimum the discrepancy between the complexity of the actual relationships, and the simplicity of those presupposed by ordinary grammatical forms, by speaking of the observed somatic change as the reaction of the organism to the gene substitution in question. We should then at least avoid any impression of vagueness or contradiction if differently constituted organisms should be found to react differently. It is, once the matter is viewed thus, far from inconceivable that an organism should evolve, if so required, in such a way as to modify its reaction to any particular gene substitution.

There are several cases in which such modification has been observed to occur in experimental stocks. It has been rather frequently observed, when a new and sharply distinct mutant in *Drosophila* has been put aside to breed in stock bottles for some generations, that when it is required again for use, the mutant form appears to be appreciably less distinct from the wild type than it had at first seemed. The reality of this tendency to revert to the wild form, as well as its cause, has been demonstrated in several cases by the simple but crucial experiment of mating the modified mutants to unrelated wild stock, and, from the hybrid, extracting the mutant form by inbreeding. The mutant form so recovered is found to have regained much of its original intensity; and thus shows that the modification has not been due to any change in the mutant gene, but to a change in the genetic complex of the organism with which it reacts. This change is now open to a simple explanation. The flies from which the stock was formed were variable in genetic qualities which affected the violence of their reaction to the mutant gene. In the competitive conditions of the stock bottle those hereditary units which favoured a mild reaction produced flies less defective than their competitors, and the selection of these modifying factors rapidly modified the average intensity of the reaction to the mutant gene, and consequently its average divergence in appearance from the wild fly. A similar case of the partial recovery of a mutation in the nasturtium, handicapped by partial sterility, has been observed by Professor Weiss; and Mr. E. B. Ford informs me that the mutant types found in the shrimp, *Gammarus chevreuxi*, have frequently made in culture a noticeable improvement in viability. The effect of preserving a mutation in a number of individuals breeding preferentially from the least defective, is thus to modify the organism in such a way as to mitigate the disadvantageous effects of the mutation. It is not only the frequency of a gene, but the reaction of the organism to it, which is at the mercy of Natural Selection. To understand the effect of a gene on members of a given population, that is, the reaction of such organisms to it, we must consider what part that gene has played in their ancestry.

The great majority, if not all, of the mutations which we can hope to observe in experimental culture must, unless these mutations can be ascribed to our cultural methods, have occurred in the history of the species in enormous numbers: many of the *Drosophila* mutations have occurred repeatedly in culture, and, large as the numbers observed have been, they are trifling compared to the total ancestry of any individual wild fly. Our knowledge of the frequency of individual mutations is at present slender; but it is sufficient to establish that many mutations must occur with a frequency of 1 in 100,000, or 1 in 1,000,000; and, indeed, the probability of mutations much rarer than this appearing in cultures is extremely small. We have, of course, no direct knowledge of the mutation rates prevalent in nature, but what has been discovered so far of the causes affecting mutation rate gives no ground for supposing that they are lower than in the laboratory. As to the extent of the ancestry of an individual fly over which a given mutation has been liable to occur, we have good grounds for assuming that it may often be longer than the separate existence of specific types; for different species of *Drosophila* have shown several mutations which can be identified by hybridization. Beyond this, direct tests of identity fail us; but it is not an unreasonable conjecture that such a mutation as albinism, which appears in mammals of the most diverse orders, has been occurring in the ancestry of the group from its earliest beginnings. On the other hand, as will be seen below, we have reason for believing that, with the evolution of new species, new mutations do sometimes commence to occur, or at least to occur with appreciable frequency.

Modification of the heterozygote

When an unfavourable mutation persists in occurring in every generation once, let us say, in each million chromosomes, it will, of course, be kept rare by selection; but it will, on the other hand, affect many individuals who are potential ancestors of future generations, in addition to those in which the mutation actually occurs. An important consequence of its rarity is that both these classes will be heterozygotes far more frequently than they will be homozygotes. If p is the relative frequency in the population of mutant to wild-type genes, the three classes of individuals, non-mutant, heterozygote and homozygous mutant, will appear in the ratio $1:2p:p^2$, so that even if p were as large as one-thousandth, the heterozygotes would be 2,000 times as numerous as the mutant homozygotes. A consequence of this is that, so long as the heterozygote differs from the wild type appreciably in fitness to survive, the relative numbers of the three classes will be determined, for a given mutation rate, by the selective advantage of the heterozygote, and to no appreciable extent by the selective disadvantage, or even complete lethality, of the mutant homozygote.

For our present purpose we take as the relative fitness of the heterozygote, denoted by v, the ratio which the average number of offspring of this type bears to the average from non-mutant individuals. Then it is easy to see that the fraction p will be diminished in each generation by the quantity $p(1-v)$ and, so long as p is small, will be augmented by the quantity k representing the actual mutation rate. An equilibrium will therefore be established between the agencies of mutation and selection when

$$p(1 - v) = k.$$

If, to take one extreme, v is a small fraction, then p is little greater than k, little greater, for example, than 1 in 1,000,000, and at this extreme the heterozygotes will occur 2,000,000 times as frequently as the mutant homozygotes. If v is $\frac{1}{2}$, p will be twice k, and the heterozygotes will still be a million times the more frequent. If on the other hand the viability and general fitness of the heterozygotes are so good that it is only at a 1 per cent. disadvantage, and $v = 0.99$, the heterozygotes will still be 20,000 times the more frequent.

These very high ratios justify the conclusion that if the heterozygote is at any appreciable disadvantage compared to the wild type, it will be so enormously more frequent than the homozygote that any selection of modifiers which is in progress will be determined by the reaction of the heterozygote.

Two other circumstances serve to increase the disproportion of the selective effects. In the first place, the efficacy of the selection in modifying the characteristics of the species depends not only upon the frequency of the individuals selected, but upon their chance of leaving a remote posterity. In fact we need to evaluate not the relative numbers of the two types in any one generation, but the proportions they represent of the total ancestry of a distant subsequent generation. Evidently, if, as is to be anticipated, the viability of the homozygous mutant is lower than that of the heterozygote, the latter will count for more in future generations, and even if the two types had equal viability, the heterozygote is still at an advantage, for mated with wild type only half his offspring will be heterozygous, while in a similar case all the offspring of the homozygote will be equally handicapped.

This point becomes of importance with sex-linked factors, where the mutant type males and the heterozygous females do not differ greatly in frequency, but may differ greatly in viability, with the result that the latter may occur much more frequently in the ancestry of the existing wild population.

In the second place, on any biochemical view of the intracellular activity of the genes, it is difficult not to admit the probability that the heterozygote may be inherently more modifiable than are the two homozygotes, especially in respect to the differences which distinguish these last; for in modifying the effect of the homozygote we must imagine the modifying gene to take part in some reaction which accentuates or inhibits the effect in question, while in the heterozygote the original ingredients are already present for all that normally takes place in the two corresponding homozygotes.

Since the first edition of this book it has become recognized as an almost invariable rule, when the three genotypes can be compared, that the wild type is found to be the least variable, the heterozygote the most so, while the homozygous mutant occupies in this respect an intermediate position.

The fraction of the ancestry of future generations, ascribable to heterozygotes, though greatly exceeding that due to mutant homozygotes, is still absolutely small. We may obtain the proportion ascribable to a single heterozygote, compared to a non-mutant, by equating it to half the proportion ascribable to its probable offspring: thus if the proportions due to heterozygotes and non-mutants are as $x : 1$ we shall have

or

$$x = \frac{1}{2} v \left(1 + x\right),$$

$$x = \frac{v}{2 - v} \, ;$$

and, since the proportion of the population which is heterozygous is

$$\frac{2k}{1 - v} \, ,$$

their proportionate contribution to remote future generations is found to be

$$\frac{2 \, kv}{(1 - v)(2 - v)} \, .$$

This quantity which, when the mutation rate (k) is 1 in 1,000,000, rises to about 1 in 5,000 if v is 0·99, represents the rate of progress in the modification of the heterozygote, compared to the rate of progress which would be effected by selection of the same intensity, acting upon a population entirely composed of heterozygotes. In the case of homozygotes the progress made by the Natural Selection of modifying factors has been shown to be far from negligible, even over short periods of observation, and under the serious restriction that the supply of modificatory variance is limited by the small number of the original stock. In considering the modification due to the selection of heterozygotes in nature, we may fairly assume that these are at least as liable to genetic modification as are homozygous mutants, and that a selection acting only on 1 in five or ten thousand of the population will have no appreciable influence in reducing the variance available.

Special applications of the theory

At the time of the first edition the reactions of the Crinkled Dwarf mutation in the cultivated New World cottons were thought to provide clear and simple evidence of the evolution of dominance comparatively rapidly. Further information on this case shows it to be quite complex, and that while Hutchinson was undoubtedly right, in what is now generally agreed, that modifying factors are abundant, and that much modification has taken place in the phenotypic appearances of the different hetero-zygotes and homozygotes at this locus, yet these are more numerous than was at first thought, and the interpretation at first put on some of the facts is untenable.

At the present time there is no need to rely on the indirect evidence of such cases. In 1958 it seems more useful to give a short account of some of the direct experimental tests which have been made. They demonstrate the ease with which selection is able to modify the heterozygote, relative to the two homozygotes (though these may also be somewhat affected), so as to change or reverse the relationship of dominance. The first of these, and in some ways the most important, was carried out by Dr. E. B. Ford at Oxford with the Currant Moth (*Abraxas grossulariata*).

Ford chose, as suitable for a decisive test, the factor responsible in the heterozygote for the variety *semi-lutea* Raynor, and when homozygous for the variety *lutea* Cockerell. In the latter the wings, instead of being a pale cream spotted with black, are suffused with a deep sulphur yellow. The heterozygote is intermediate, favouring somewhat the normal form, so that in Nature the mutation is somewhat more recessive than dominant; both varieties are, however, appreciably variable. Ford established for comparison with the moths bred a colour scale of eight grades, of which number one was white and number eight of the darkest tint exhibited by specimens of *lutea*. In terms of these grades the results of his matings may be expressed very simply. By interbreeding heterozygotes the three genotypes were produced in each generation in the ratio 1:2:1. Starting with a heterozygote obtained in a culture of wild larvae, the experiment consisted in interbreeding in the first generation unselected heterozygotes, and in subsequent generations selecting moths judged to be heterozygous in two lines, one from among the darkest (E, F and G), and the other in the negative direction from among the lightest (B, C and D). For three generations, therefore, selection was practised in opposite directions.

The effect of selection was immediate and striking. From being trimodal with an inclination to recessiveness in the first (unselected) generation, in B it is already bimodal, with confluence of the normals and heterozygotes, and in stage D the gene is frankly recessive. Concurrently, selection in the positive direction shifted the mode of the heterozygotes in each generation, so that in the third generation it is confluent with that of the homozygous *lutea*. At this stage the mutation was effectively dominant, though doubtless the average depth of pigmentation of the two overlapping genotypes was still appreciable.

The course of the experiment is shown in the figure (Fig. A) reproduced from *The Annals of Eugenics* by kind permission of Professor L. Penrose. Examining the diagrams carefully, it will be seen that the heterozygote is at all stages the most variable form, while the mutant *lutea* is considerably more variable than the typical genotype. Such a specific contrast is not easily ascribable to the peculiarities of the metric employed, for the same classes, principally numbers 5 and 6, are occupied by heterozygotes in stage G, but by homozygous *lutea* in stage D; while the normals of stage G are distributed very much as must be the heterozygotes of stage D; and in these comparisons also the heterozygotes are equally the more variable.

The important point demonstrated by this experiment is that animals drawn directly from the wild population contain such an abundance of auxiliary heritable factors capable of modifying the dominance of the factor chosen, that in no more than three generations' selection it can be changed in one direction to a perfect recessive, and in the other to a somewhat less perfect dominant. There is no reason for expecting that modifiers are less available for any other factor, had selection been directed towards accumulating their effects. In this and in similar experiments, continuation of the process encounters the difficulty that, when there is much overlapping between the heterozygote and one of the homozygous forms, many insects selected for breeding as heterozygotes will turn out to be in reality homozygous, and their broods will be useless for the purpose intended. It is, however, sufficient to demonstrate that modifying factors are available, in a stock chosen blindly, to shift the appearance of the heterozygote from that of one homozygote to that of the other, for this proves that the reluctance at first felt to postulating the normal existence of such factors was due to an understandable tendency to underrate the potentialities of residual inheritance so long as it has no visible manifestation.

That the response to selection found by Ford with the Currant Moth was not in any way exceptional was further shown by an experiment (Fisher and Holt 1944) carried out with mice in the years 1940–41. The writer had bred a number of partly inbred lines of mice for some years prior to 1940, but did not possess a particular mutant known as Danforth's short tail (*Sd*), which had in the meanwhile been discovered in the United States. In 1940, by the courtesy of Professor L. C. Dunn of Columbia University, some specimens of these were obtained in his colony.

In the heterozygote the effect of the mutant is to shorten the tail, and often to introduce sharp bends, as with the bulldog. The delicately graded series of vertebrae towards the tip are completely absent. In the material received the tail was about one third of the normal length. The homozygotes were tailless and, according to Dunn, never survived for more than 24 hours. They usually or always lacked kidneys, and had characteristically an imperforate anus. The question to be determined experimentally was whether my own stock, in which this mutant had never appeared, contained invisibly a number of modifying factors capable of influencing the length of the tail developed by these heterozygotes.

Since the minimal length of the generation in mice, about two months, is much less than their fertile life, which may exceed twelve months, the experimental procedure was different from Ford's, although, as in his case, selected heterozygotes were interbred in each generation. At intervals of about two months a survey was made of young stock bred since the last survey, and selected pairs were set aside for a positive and for a negative selection line. Matings, unless terminated prematurely, were maintained for eight months. Usually the longest and shortest tailed of those available were bred from, though we were occasionally influenced by the record or by the health and fecundity of the different matings.

The imported mice were crossed, of necessity blindly, to mice of my own stocks, and the crossed mice interbred to eliminate albinism, for the imported mice were albinos. Great variability in tail length appeared at this stage. That this variability had a genetic basis was shown by the output of the first selected matings. Indeed within six months the two lines were completely separate.

In subsequent selections during two years in all, the positive line advanced continually to an average length of about 50 mm., at which point many heterozygotes could only be recognized to be not normal by a minute examination, and some were beginning to be classified as normal. At this stage the mutation was nearly recessive. In the negative line an average tail length of about 12 mm. was quickly established, but this did not easily decrease further, by reason of natural selection coming into play in the opposite direction and eliminating a proportion of the mice more seriously affected. Elimination in the nest between birth and separation from the parents is unavoidable, and it was shown that selective elimination was active in the negative line, and in unselected mice, but not in the positively selected line.

Previous work by Professor L. C. Dunn and his associates had shown that when constantly backcrossed to a certain inbred strain, the Bagg albino, the effect of the *Sd* factor was strongly enhanced, so that many heterozygotes were found to show the lethal defects characteristic of homozygotes. The counterpart of this observation was found in our own work, in which a number of homozygotes were found not to be wholly lethal, but in some cases to survive beyond weaning. Dissection has shown very abnormal internal development, but sufficient kidney tissue evidently to support life, apparently normal ovaries, and development of one oviduct in a female apparently competent for reproduction. I have in no case succeeded in breeding from one of these animals, and the additional strain of pregnancy would presumably be fatal, although there is no reason to suppose that in no case could they possibly breed. Naturally, they occur only in the positive selection line.

A group of facts of very particular interest in this connexion is presented by domestic poultry. Crosses between the different breeds show that many of the distinctive breed characteristics are due to simple Mendelian factors. In a number of cases, however, it is the fancy breed character, and not the character of the wild *Gallus bankiva*, which is found to be dominant. There must be a dozen or more factors of this kind; three are known which affect the conformation of the comb; one produces a crest; there is a dominant white which inhibits pigment formation in the plumage; and others influencing the colour or pattern of the feathers, or the colour of the shanks. Domestic poultry show also mutants of the kinds familiar in other organisms, recessives and lethal 'dominants', but they are peculiar in this surprising group of factors which are non-lethal and dominant to the wild type. It is noteworthy that none of these factors originated in a recorded mutant and that their effects, while presumably they would be deleterious in the wild environment, are not pathological in the sense of impairing the vitality of the birds as domestic poultry. They are all, in fact, definite breed characteristics. Other birds bred in captivity seem to have thrown the ordinary excess of recessive mutants, such as cinnamon canaries, or yellow budgerigars; and for each of these reasons we should be led to seek for an explanation of the peculiarity of the domestic fowl rather in the conditions of its domestication than in the nature or environment of the wild species. In the former there seems to be one very striking circumstance which throws light on the dominant characters of the domestic breed.

The wild jungle fowl is common in many parts of India, and it has frequently been observed that the wild cocks mate, when opportunity is afforded, with the hens of domestic flocks. If this is so down to the present day, we may infer that it has been so since the earliest stages of domestication, and indeed that it was the prevalent condition throughout the period, probably a long one, when the fowl was only kept by jungle tribes. I do not postulate that the cocks were not kept; for they may have been valued for cock-fighting as early as the hens for egg-production; moreover, some of the factors concerned are sex-linked, and would only show dominance in the cock; but it is probable, and indeed almost impossible to dispute, that for long ages the domestic flocks were continually liable to be sired by wild birds. In the case of most domestic animals and plants, recessive mutations, when they appear, will immediately breed true, and man's curiosity and love of novelty have thus repeatedly led him to perpetuate forms which, as often as they appear in a state of nature, are eliminated by Natural Selection. On crossing with the wild form such recessive characters disappear and seem to be lost, and if such crossing is at all frequent, the only mutations which could lead to constant breed characteristics would be those that were not completely recessive. With these some of the chicks would always show the breed characteristic, and a continued selection or preservation of the valued types would retain their character in the breed. Moreover, since these types are only to be retained by selection, it is certain that selection would favour those individuals in which the mutant characteristic reached the most pronounced development. Man, in fact, whenever his broods consisted half of heterozygotes and half of wild-type fowls, if he valued the heterozygote characteristics, and therefore selected them rather than the others, would also, necessarily, at the same time select those heterozygotes in which the mutant gene was least recessive or most dominant.

It will be noticed that on this view of the origin of some of the breed characteristics of the domestic fowl, we have an explanation of two distinct peculiarities which these characters exhibit; namely both the high proportion of mutant characters which are not recessive to the wild type; and of the high degree in which dominance is developed, at least in certain breed crosses. It is important, too, in this connexion, that other crosses are known in the case of several of these factors, in which dominance appears to be incomplete. A full and satisfactory examination of such cases would seem to be possible only by introducing the mutant gene, and very little else, into breeds in which this gene is unknown; for dominance can only properly be examined if the two homozygotes and the heterozygote have, in other respects, a similar genetic composition. It may be mentioned that my inference concerning the modification of dominance in mutant factors in the fowl, is open to the crucial test of introducing one or more of these dominants into a genuinely wild strain of jungle fowl. If my inference is correct, the mutant would then be found to be clearly intermediate, and not either completely dominant or completely recessive. Through the kindness of the Zoological Society of London, and the generosity of Mr. Spedan Lewis, it has been possible to start this experiment; the result cannot, of course, be known for several years.

The experiment was carried out and completed in the seven years 1929–35. Of the 'dominants' used, Crest, Black internal pigment, Rose comb, Feathered feet, and Polydactyly were obtained from birds of the Japanese Silky breed; Pile (dominant white) and Barred were from White Leghorns. Each dominant, in what became seven separate lines, was bred five times into wild stock, and after the fifth 'introgressive hybridization' heterozygotes were interbred in order to obtain homozygotes, and to compare these with the heterozygotes, on a background of wild germ-plasm. The experiment has been reported fully in 1935 in *Phil. Trans.* B, 225:195–226, and in 1938 in *Proc. Roy. Soc.* B, 125:25–48.

In brief it may be stated that with the possible exception of Rose comb, on which my data were incomplete, none of these factors showed complete dominance. Pile was most nearly a dominant, though even here the differences between the homozygous and the heterozygous birds were clear. Feathered feet was very nearly recessive; Polydactyly, Barred and Black internal pigment gave easily distinguishable heterozygotes. In the case of the barred factor, which is sex-linked, the difference between homozygotes and heterozygotes is the difference on which sex discrimination is now based in the so-called auto-sexing breeds, for the apparent dominance of the factor as used by the earlier geneticists disappears when a secondary factor for Black is removed.

The case of Crest was particularly instructive, for it was found when homozygous (in the wild stock) to give a simple cerebral hernia, such as had been observed and recognized as a simple recessive in the earlier work with poultry (*Science*, 80:288–9, 1934). In fact Crest should never have been described as a dominant, but at most as the partially dominant effect of a factor which also had recessive effects. Since crested breeds such as the Silky and the Polish must be principally homozygous for this factor, and they are not characteristically herniated, though, in the case of the Polish, having a peculiar domed and fenestrated skull, it appears that in these breeds, during their domesticated history, the injurious hernia has been repaired more or less elaborately in such a way that the desired character of the crest is maintained and enhanced, while the undesired hernia is suppressed. In the case of the Silky this suppression is so complete that the mutant gene might properly be called a dominant, so great is the difference from the wild species. In the wild stock indeed the whole group of 'dominants' tested are in their reactions quite comparable with the semi-dominant mutations often observed in o her experimental species.

The process of modification

The case of fowls confirms, so far as it goes, the other evidence available as to the speed with which dominance may be modified; for in this case, although the whole process has perhaps occupied no more than a thousand generations, the effective selection is applied, not to a population containing only one heterozygote in 10,000 or so, but to broods half of which are heterozygotes; and moreover in which *ex hypothesi* it is the heterozygotes rather than the wild type that are chosen to continue the breed. Evolution under such human selection should, therefore, take place many thousand times more rapidly than the corresponding evolution of recessiveness in nature.

As to the speed of the latter process, the principal unknown element for a mutation of given viability (v) and mutation rate (k) is the quantity of modificatory variance available to influence the heterozygote. This will presumably tend nearly to zero as v tends to unity, but its relation to v for values differing considerably from unity will be somewhat different according to the different views which we may form as to the manner in which the modification is brought about. In the case of homozygotes we must suppose that the modifying factors, by intensifying the appropriate developmental reactions, succeed, in effect, in remedying the situation which arises at that stage at which defective development is initiated. This may also be true of heterozygotes, and, if the greater part of the modificatory variance available is of this sort, we should expect its magnitude, *ceteris paribus*, to depend only upon v, and consequently that all mutations would follow one another along the same path towards

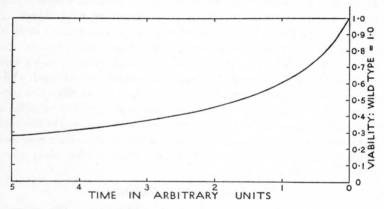

FIG. 4. The relation between the severity of the handicap imposed by a mutation, and the time needed to repair the defect by the selection of modifiers, supposing the variance of v to be proportional to $v (1-v)$.

normality at speeds proportional to their mutation rates, but otherwise dependent only on the stage which they have at any moment reached. Such a view is illustrated in Fig. 4.

On the other hand it does not seem, in the present state of knowledge, improbable that the greater part of the variance may be due to a cause special to heterozygotes; namely the varying extent to which one or other of the homologous genes may be allowed to take part in the nuclear reactions for which they are responsible. On this view the amount of variance available would depend, not only on the viability actually attained, but upon its original value; being, for heterozygotes of the same viability, greater for mutations having the larger effect. We should then obtain such a series of trajectories as is illustrated in Fig. 5.

In either case the final stages of approach to normality will be the most rapid, and a mutation which makes a bad start may have made but little progress by the time other mutations, which have occurred no more frequently, have attained complete normality. We should of course expect to find most cases at the stages where progress is slowest, and a comparatively large accumulation in any stationary condition. The relatively rare 'dominant' mutants of *Drosophila* may be regarded either as comparatively new mutations, or more probably, as regards the greater number of them, as mutations in

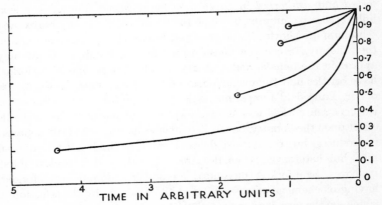

FIG. 5. Trajectories of improvement of the heterozygote, on the supposition that the modificatory variance depends also on the magnitude of the unmodified effect.

which the heterozygote has been throughout its history so severely handicapped, that little progress has been made. The greater number of observed mutations are found, as would be expected, in the resting stage of complete recessiveness, and in the case of the lethals, whose condition should be absolutely stationary, the number accumulated is enormous. With non-lethal mutants, after the heterozygote has become, within a very minute difference in viability, equivalent to the wild type, a process of modification of the homozygote may be expected to commence; and this for the same level of viability, should, on the view that the homozygote is not much less modifiable than the heterozygote, be comparable in speed with the modification of the latter. The second process would, apart from any difference of modifiability, presumably be appreciably slower than the first, for the homozygote may be expected to be initially much the more heavily handicapped, though its viability may have been, incidentally, considerably improved during the process of modification of the heterozygote. Nevertheless, we must be prepared to admit that innumerable mutations may have occurred in the past, of which even the homozygote has become to all appearances normal, and which consequently leave no trace for genetic research to reveal. There appears to be no reason, however, why such factors should not function in special cases in modifying the effects of rare mutants.

A case of interest in this connexion is presented by the two factors *forked* and *semiforked* in *Drosophila melanogaster*. Forked is a sex-linked recessive mutant, in which the bristles of the head, thorax, and scutellum are shortened, twisted and heavier in appearance, than in the wild fly. Since the factor is sex-linked, dominance can only appear in the female, and ordinary females heterozygous for forked have bristles indistinguishable from those of wild flies. In the course of Dr. Lancefield's experiments with this factor in 1918, however, the gene semiforked was discovered; this gene has no distinguishable effect upon the homozygous forked females, or upon the forked males; it produces, but rarely, a slight shortening of the bristles in normal flies, but heterozygous females are modified by it into clear intermediates. Semiforked thus acts as a modifier of dominance in forked, having biochemical effects similar to those factors by the selection of which, on the view here put forward, its dominance has been acquired. It is, however, scarcely probable that semiforked is actually one of these factors, for it is itself a recessive, as judged by its interaction with heterozygous forked. It may, on the other hand, well be an old mutation which has reached a stage in modification at which even when homozygous it exerts scarcely any effect.

Inferences from the theory of the evolution of dominance

One inference that may fairly be drawn from the foregoing considerations is that the widely observed fact that mutations are usually recessive should not lead us to assume that this is true of mutations of a beneficial or neutral character. On the contrary, we have reason to believe that it is confined to a class of mutation which persistently recurs, with a mutation rate not greatly less than one in a million, and which has been eliminated with equal persistence by Natural Selection for many thousands, or possibly millions of generations. This class of mutation is, and will perhaps always be, of the greatest value to the plant-breeder and the geneticist, for it supplies them with their most prized variants, but we have no right on this account to suppose that it has any special importance in evolution. With mutations not of this class we have no reason to expect dominance in either direction. *A priori* it would be reasonable to suppose that at the first appearance of a mutation, the reaction of the heterozygote would be controlled equally by the chemical activity of the two homologous genes, and that this would generally, though not necessarily in every individual case, lead to a heterozygote somatically intermediate between the two homozygotes. We should of course not expect all quantitative differences to be numerically equal, for these depend upon our methods of measurement, and to take a simple analogy, the removal of half the pigment from a black structure might well be judged to produce less effect than the removal of the remainder. To postulate equal functional importance of the two homologous genes is therefore not to assist the impossibility of all appearance of dominance, but that a general intermediacy of character, such as that to which attention has already been called in heterozygotes between different mutants of the same gene, should be the prevalent condition. The change brought about in a species by the acquisition of a favourable mutation will thus generally take place by two not very unequal steps taken successively in the same direction. It is interesting that this situation bears some slight resemblance to the successive mutations in the same direction, imagined by Weismann.

The case of the evolution of dominance serves to illustrate two features of Natural Selection which, in spite of the efforts of Darwin, still constitute a difficulty to the understanding of the theory, when the latter is illustrated by the active care of the human breeder in selecting his stock. These are the absence of any intention by nature to improve the race, and the fact that all modifications which tend to increase representation in future generations, however indirectly they may seem to act, and with whatever difficulty their action may be recognized, are *ipso facto*, naturally selected. The acquirement of dominance to harmful mutants cannot properly be said to improve the species, for its consequence is that the harmful genes are concealed and allowed to increase. There is some analogy here with Darwin's theory of sexual selection in so far as this is applied to characters of no use to the organism in relation to its environment or to other organisms, and to qualities which do not assist the sexes to discover and unite with one another, but only to qualities which are preferred by the opposite sex. Even in such cases, however, sexual selection does give a real advantage to one half of the species in relation to one situation of their life-history, while in the selection of dominance the genetic modification of the whole species results in the structural modification of an incomparably smaller fraction. If we adopt Darwin's analogy of a human or super-human breeder scrutinizing every individual for the possibility of some direct or indirect advantage, the case of the evolution of dominance shows well how meticulous we must imagine such scrutiny to be.

We have seen in the previous chapter, in general terms, that the difficulty of effecting any improvement in an organism depends on the extent or degree to which it is adapted to its natural situation. The difficulties which Natural Selection has to overcome are in this sense of its own creating, for the more powerfully it acts the more minute and intricate will be the alterations upon which further improvements depend. The fact that organisms do not change rapidly might in theory be interpreted as due either to the feebleness of selection or to the intensity of adaptation, including the complexity of the relations between the organism and its surroundings. We have no direct measure of either value, and the point at issue can only be expressed in concrete terms in relation to some definite change, real or imaginary, in some particular organism. For this purpose the recessive mutations seem to supply what is wanted, and the reader who accepts the conclusions of this chapter will perceive that any maladaptation of the same order of magnitude as these, and equally capable of modification, would be remedied by Natural

Selection some ten or hundred thousand times more rapidly than dominance has actually been acquired. To take a more real case, instead of imagining that a whole species were suddenly changed so as to be as ill-adapted to its conditions as our familiar mutants, if we suppose that the organic or inorganic environment of the species were to change suddenly, or that a colony of the species were to find itself in surroundings to which it was equally ill-adapted, we have analogous reason to suppose that the evolution of adaptive characters would proceed at the same rate. It might indeed be said that each mutation is such an experiment in little.

With regard to the precision with which adaptation is in fact effected we must be careful to remember that all of the heterozygotes of the different recessive mutations, including, apparently, thousands of recessive lethals, are genetically different. However indistinguishable the end products may be, these are produced by different developmental processes, even if the ultimate differences are only intracellular reactions. The fact that they are much alike can only be interpreted as showing that likeness of this degree is requisite, even for such approximately normal adaptation as is required of a rather rare heterozygote. If any appreciable diversity of form were possible within the range of such approximately equal adaptation we should surely find it among this multitude of heterozygotes. Since any differences which may exist between them are certainly extremely minute we have here a clear indication of the closeness with which any tolerably successful individual must approach the specific type, and an upper limit of the magnitude of the differences, which have a reasonable chance of effecting improvement.

Summary

Examination of the incidence of dominance in mutations observed to occur, and of other genes which must be regarded as mutants, shows that in the majority of cases the wild gene is dominant to the mutant genes, while in a minority of cases dominance is incomplete. Different mutations of the same wild genes show mutually on the other hand a regular absence of dominance. If the substitution of mutant for primitive genes has played any part in evolution these observations require that the wild allelomorphs must *become* dominant to their unsuccessful competitors.

The incidence of heterozygotes of each mutant among the ancestry of the wild population is, if we may rely upon observed mutation rates to be of the right order of magnitude, sufficient to account for the evolution of dominance by the selection of modifying factors. This process is extremely slow, since the proportion of the population effectively exposed to selection is only about 1 in 10,000 or 100,000.

A number of experimental trials have now demonstrated that dominance can be modified more readily than was at first thought possible; the anomalous occurrence of dominance in domestic poultry may be interpreted as due to the effects of human selection in flocks liable to be sired by wild birds.

The theory of the evolution of dominance thus accounts for a considerable body of facts which have received so far no alternative explanation. If it is accepted it appears to throw considerable light upon the nature of mutations, and on the intensity of adaptation; in particular the closeness of the convergence of very numerous heterozygous genotypes indicates somewhat forcibly that adaptive significance, sufficient to govern evolutionary change, is to be found in differences of much less than specific value.

REFERENCES

1. R. A. Fisher (1934). 'Crest and hernia in fowls due to a single gene without dominance.' *Science*, 80:288-9.
2. R. A. Fisher (1935). 'Dominance in poultry.' *Phil. Trans.*, B, 225:195-226.
3. R. A. Fisher (1938). 'Dominance in poultry. Feathered feet, Rose comb, Internal Pigment and Pile.' *Proc. Roy. Soc.*, B, 125:25-48.
4. R. A. Fisher and S. B. Holt (1944). 'The experimental modification of dominance in Danforth's short-tailed mutant mice.' *Ann. Eugen.*, 12:102-20.
5. E. B. Ford (1940). 'Genetic research in the Lepidoptera' (Galton Lecture). *Ann. Eugen.*, 10:227-52.

IV

VARIATION AS DETERMINED BY MUTATION AND SELECTION

The measurement of gene frequency. The chance of survival of an individual gene; relation to Poisson series. Low mutation rates of beneficial mutations. Single origins not improbable. Distribution of gene ratios in factors contributing to the variance. Slight effects of random survival. The number of the factors contributing to the variance. Chapter V. The observed connexion between variability and abundance. Stable gene ratios. Equilibrium involving two factors. Simple metrical characters. Meristic characters. Biometrical effects of recent selection. Summary.

> *There was a first occurrence, once for all,*
> *Of everything that had not yet occurred.*
> SOPHOCLES.

The measurement of gene frequency

In Chapter II considerable emphasis was laid on the fact that the heritable variance displayed by any interbreeding group of organisms has no inherent tendency to diminish by interbreeding, provided that the variance is due to differences between particulate genes, which segregate intact from all the genetic combinations into which they may enter. In such a system any changes in variability which may be in progress must be ascribed to changes in frequency, including origination and extinction, of the different kinds of genes. In the present chapter we have to inquire into the causes which determine the degree of variability manifested, or in other words, into the level of variability at which the origination and extinction of genes are equally frequent.

It will be sufficient at first to treat in detail the case of dimorphic factors, that is, of loci to occupy which there are only two kinds of genes available. Cases of three or more alleles, and the important extension to polysomic inheritance, can be elucidated by the same methods, without introducing any new principle. In considering dimorphic factors we shall be concerned with the relative frequency of only two kinds of genes, which we have represented in previous chapters by the ratio $p : q$, and with the causes by which their frequencies are modified. It is therefore of some importance to adopt an appropriate scale on which such changes of the frequency ratio may be numerically measured. It would of course be possible to adopt a percentage scale for such measurement, to distinguish factors according to the percentages of the loci available occupied by the two types of genes. We should thus distinguish factors in which each type of gene occupied 50 per cent. of the loci available, from factors in which the more numerous type of gene occupied 60 or 90 or 99 per cent., and discuss with what frequency factors might be expected to lie in the regions bounded by these values; what proportions of the factors, that is to say, should be expected to have their more numerous gene occupying between 50 and 60 per cent. of the loci, what proportion between 60 and 70 per cent. and so on. In cases where dominance has been developed we might ask the same questions respecting the frequency not of the more frequent, but of the dominant gene; and would thus distinguish cases in which from 20 to 30 per cent. of the genes were dominants, from cases in which the proportion lay between 70 and 80 per cent. For all purposes of this kind, however, in view of the actual relationships to be discussed it is more useful to use a scale on which the ratio between the two frequencies increases in geometric progression. Starting from the case in which the two frequencies are equal and each gene occupies 50 per cent. of the available loci, we should then regard the frequency ratios 2 : 1, 4 : 1, 8 : 1, 16 : 1 and so on, as equal steps of increasing frequency, although the corresponding percentages are 66·7, 80·0, 88·9, 94·1. Such a scale is symmetrical. If we step off in the opposite direction we shall arrive at the frequency ratios 1 : 2, 1 : 4, 1 : 8, 1 : 16 with the complementary percentages. Mathematically the scale we have chosen is equivalent to measuring the frequency ratio by the variate

$$z = \log\frac{p}{q} = \log p - \log q.$$

If the logarithms are taken to the base 2, our steps will be each of unit length, while if we use, as is mathematically more convenient, natural or Naperian logarithms, the steps, while still being of equal length, will be about 0·7 of a unit. The two practical advantages of the use of the logarithmic scale for the frequency ratios of a dimorphic factor are, firstly, that it enables an adequate distinction to be drawn between the very high frequency ratios such as a thousand million to one, which can occur in the genes of numerous species, and more moderate frequency ratios such as 1,000 to 1 which are almost indistinguishable from them on a percentage scale; and secondly, that the effects of selection in modifying the gene frequencies are, on the logarithmic scale, exhibited with the utmost simplicity, namely by changes of position with velocities that are uniform and proportional only to the intensity of selection.

For factors which are not sex-linked, each individual will contain two genes like or unlike each other. If every individual in a species is thus enumerated, and counted as two, the maximum attainable frequency ratio will be effectively the ratio which twice the number of individuals in the species bears to unity. The range of possible frequency ratios on the logarithmic scale thus depends on the number of individuals in the species, and it is easy to see that it is increased by 2 log 10, or 4·6, if the population in the species is increased tenfold. For example, a species of 10,000,000,000 individuals will give a range of values from about − 23·7 to + 23·7. Of this range about 5 units at either end represent cases in which the less frequent gene exists in only about 100 or less distinct individuals, or to be more exact, since 1 individual can contain 2 such genes, in about 100 homologous loci. In these regions it is clear that the rarer gene is, relatively speaking, in some danger of extinction, and the absolute length of these regions on our scale will not depend on the number of individuals in the species. Between these two extreme regions lies a central region in which both genes are comparatively numerous, at least in the sense that neither of them will exist in less than 100 individuals. It is the length of this central or safe region which depends on the magnitude of the population of the species. For 10,000,000,000 it is about 37 units in length, for 100,000,000 it has only about 28 units. The logarithmic scale thus affords a simple demonstration of the important bearing which population size has on the conservation of variance.

In the hypothetical enumeration of the genes of the population considered in the last paragraph, no account was taken of the age or reproductive value of the individuals enumerated. If account is taken of these there is no limit to the magnitude of the frequency ratio attainable in either direction, but the distinction between the relative insecurity of the rarer gene in the extreme regions, and its relative security in the central region, is still valid. This statement is based on the circumstance that a gene which exists in a dozen individuals who, in the sense of Chapter II have low reproductive value, is in at least as much danger of extinction as one existing in a single individual whose reproductive value is equivalent to that of the twelve others put together. For mature forms the probability of survival must be nearly equivalent. If, however, the reproductive value we are considering is supplied entirely by immature or larval forms, normally liable to great mortality, the chance of extinction for a given amount of reproductive value may be considerably enhanced. It is therefore convenient to exclude the immature forms altogether from discussion, and to consider the results of enumeration in which individuals are only counted when they attain to the beginning of the reproductive stage of their life history. We shall count each generation near the maximum of its reproductive value, and when its numbers are least. The magnitude of the population of a species can then be conceived, not by the analogy of a census enumeration, in which individuals of all ages are counted, down to an arbitrary legal minimum at birth, but as the number of individuals of each generation who attain to the reproductive stage. In species having several generations in the year, the numbers of which are also much affected by the annual cycle, it is probable that the conclusions to be drawn as to the effects of population size, will be most nearly applicable to the normal annual minimum of numbers.

The chance of survival of an individual gene

An individual gene carried by an organism which is mature, but has not reproduced, will reappear in the next generation in a certain number 0, 1, 2, 3 etc. of individuals or homologous loci. With bisexual organisms these must of course be separate individuals, but where self-fertilization is possible the same gene may be received by the same individual offspring in each of its two parental gametes, and if such an individual survives to maturity our original gene will thus be doubly represented. In general we shall be concerned with the total number of representations, although it will be convenient to speak as though these were always in different individuals. The probabilities that of the offspring receiving the gene, 0, 1, 2 . . . attain maturity will be denoted by

$$p_0, \; p_1, \; p_2, \; \ldots \ldots \ldots ,$$

where, since one of these contingencies must happen,

$$p_0 + p_1 + p_2 + \; \ldots \ldots \ldots = 1$$

In order to consider the chances in future generations we shall first calculate the appropriate frequencies for the case in which our gene is already represented in r individuals. In order to do this concisely we consider the mathematical function

$$f(x) = p_0 + p_1 x + p_2 x^2 + \ldots \ldots$$

This function evidently increases with x from p_0, when $x = 0$, to unity when $x = 1$. Moreover, if the r individuals reproduce independently, the chance of extinction in one generation will be p_0^r. The chance of representation by only a single gene will be

$$r p_0^{r-1} p_1,$$

and in general the chance of leaving s genes will be the coefficient of x^s in the expansion of

$$(f(x))^r.$$

Now, starting with a single gene, the chance of leaving r in the second generation is p_r, and the chance that these leave s in the third generation will be the coefficient of x^s in

$$p_r (f(x))^r.$$

It follows that the total chance of leaving s in the third generation, irrespective of the number of representatives in the second generation, will be the coefficient of x^s in

$$p_0(f(x))^0 + p_1(f(x))^1 + p_2(f(x))^2 + \ldots \ldots \ldots,$$

or, in fact, in

$$f(f(x)).$$

This new function, which is the same function of $f(x)$ as $f(x)$ is of x, therefore takes the place of $f(x)$ when we wish to consider the lapse, not of one but of two generations, and it will be evident that for three generations we have only to use $f\{f(f(x))\}$, and so on for as many generations as required.

There are good grounds for supposing that if, as has been suggested, enumeration is confined to the condition of early maturity the function $f(x)$ will always have, to a good approximation, the same mathematical form. If we consider, for example, any organism capable of giving rise to a considerable number of progeny, such as a cross-pollinated cereal plant, it appears that each sexually mature individual is the mother of a considerable number, let us say 40, mature grains, and the father, on the average, of an equal number. Into each of these grains any particular gene has an independent probability of one half of entering. But since, of these grains only 2, on an average, will be represented in next year's crop by mature plants the chance of both entering into the grain and of surviving in it is only 1 in 80. The probabilities therefore of the gene reappearing in the following year in 0, 1, 2 . . . individuals will be the coefficients of x^0, x^1, x^2, . . . in the expansion

$$\left(\frac{79}{80} + \frac{1}{80}x\right)^{80}.$$

These coefficients are already very close to the terms of the Poisson series.

$$e^{-1}\left\{1, 1, \frac{1}{2}, \frac{1}{6}, \frac{1}{24}, \ldots \ldots \right\},$$

or

$$e^{-1}\left(1, 1, \frac{1}{2!}, \frac{1}{3!}, \ldots \right),$$

and would become identical with them if the arbitrary number 80 were increased indefinitely. The departure from the Poisson series is in fact ascribable to artificial assumptions which for simplicity have been allowed to enter into the calculation. We have arbitrarily assumed that each plant produces the same number of grains, whereas in reality this number will be variable. The number of pollen grains also from each plant which enter into perfect seeds will vary, and the effect of this variability will be to change the distribution very slightly in the direction of the limiting Poisson distribution. In fact it is probable that in so far as the binomial distribution obtained above differs from the limiting form, it differs in the wrong direction, for the variability in the number of grains on different plants seems to be slightly greater than what is required in a perfect Poisson series.

The general character of the Poisson series which makes it appropriate to our problem is that it arises when a great number of individuals enjoy each a small independent chance of success; if the number of individuals and the chance of each are such that on the average c succeed, then the numbers actually succeeding in different trials will be distributed according to the series

$$e^{-c}\left(1, c, \frac{c^2}{2!}, \frac{c^3}{3!}, \ldots \ldots \right),$$

and this may generally be regarded as a good approximation to the chances of individual gametes produced by a single mature individual.

If the gene confers no selective advantage or disadvantage, c will be equal to unity; the values of p_0, p_1, p_2 . . . will be given by the Poisson series

$$e^{-1}\left(1, 1, \frac{1}{2!}, \frac{1}{3!}, \ldots \ldots\right)$$

and the function $f(x)$ takes the form

$$e^{-1}\left(1 + x + \frac{x^2}{2!} + \frac{x^3}{3!} + \ldots \ldots\right),$$

or

$$f(x) = e^{x-1}.$$

TABLE 2.

Number of Generations.	Probability of Extinction.		Difference.	Probability of Survival.	
	No Advantage.	1 per cent. Advantage.		No Advantage.	1 per cent. Advantage.
1	0·3679	0·3642	0·0037	0·6321	0·6358
3	0·6259	0·6197	0·0062	0·3741	0·3803
7	0·7905	0·7825	0·0080	0·2095	0·2175
15	0·8873	0·8783	0·0090	0·1127	0·1217
31	0·9411	0·9313	0·0098	0·0589	0·0687
63	0·9698	0·9591	0·0107	0·0302	0·0409
127	0.9847	0·9729	0·0118	0·0153	0·0271
Limit	1·0000	0·9803	0·0197	0·0000	0·0197

Moreover if the gene in question is increasing in frequency in each generation in the ratio c : 1, we shall have similarly

$$f(x) = e^{c(x-1)}.$$

Having obtained these forms for $f(x)$ we may trace the survival, multiplication or extinction of the descendants of single individual genes, by a mere repetition of the process of substituting $f(x)$ for x. Table 2 shows in the first column the number of generations which have elapsed from the starting-point, these numbers having been chosen so as to follow the course of the changes over a large number of generations, in a moderately compact table. These changes are most rapid at first, so that we have chosen successive steps of 1, 2, 4, 8 generations and tabulated the conditions reached after the total expiration of 1, 3, 7, 15, 31, 63 and 127 generations. The second column shows the probability of extinction, at each stage, for genes having no selective advantage or disadvantage. The numbers may also be read, ignoring the decimal point, as the number of cases out of

10,000 in which the descendants of the original gene will have become extinct. The proportion of extinctions in the early generations is extremely high, nearly 3 in 8 are extinguished in the first generation, and of the remaining 5, 2 have failed by the third generation. In 15 generations nearly 8 out of 9 will have failed. As we proceed extinctions become very much rarer, only 2·87 per cent. are lost between the 31st and the 63rd generation, and only 1·49 per cent. between the 63rd and the 127th when there are still 1·53 per cent. surviving. The survivals may best be followed in the 5th column, in which it will be seen that with the steps we have chosen, the number of survivors tends increasingly closely to be halved at each step; in fact when n is large the chance of survival for n generations is very nearly $2/n$.

For comparison the corresponding figures have been tabulated in adjacent columns for genes for which $c = 1·01$, and which consequently enjoy an advantage of 1 per cent.; the differences are shown in the 4th column. It will be seen that the selective advantage amounts ultimately, in the limit when n is increased indefinitely, to survival in just less than 2 per cent. of the cases originally started, and of this advantage very little is gained in the early stages where extinction is rapid. Of 10,000 mutations enjoying a 1 per cent. selective advantage, and which have already reached the stage of existence in one sexually mature individual, 3,642 will fail to transmit the advantageous gene to any descendant, whereas with no selective advantage whatever, only 3,679 will so fail. Even after 31 generations the number surviving out of 10,000 will be only 687 against an expectation of 589 where no selective advantage is enjoyed. The fact is, that a selective advantage of the order of 1 per cent., though amply powerful enough to bring about its evolutionary consequences with the utmost regularity and precision when numbers of individuals of the order of 1,000,000 are affected, is almost inoperative in comparison to random or chance survival, when only a few individuals are in question. A mutation, even if favourable, will have only a very small chance of establishing itself in the species if it occurs once only. If its selective advantage is only 1 per cent. it may well have to occur 50 times, but scarcely in mature individuals as many as 250 times, before it establishes itself in a sufficient number of individuals for its future prospects to be secure.

The fact that a mutation conferring an advantage of 1 per cent.

in survival has itself a chance of about 1 in 50 of establishing itself and sweeping over the entire species, shows that such mutations cannot occur with any great total frequency before this event is realized, or at least rendered certain, by the initial success of one of their number. The odds are over 100 to 1 against the first 250 mutations of such a favourable type all perishing. Consequently the success of such a mutation must become established at a time when the mutation rate of the mutation in question is extremely low, for in a species in which 1,000,000,000 come in each generation to maturity, a mutation rate of 1 in a thousand million will produce one mutant in every generation, and thus establish the superiority of the new type in less than 250 generations, and quite probably in less than 10, from the first occurrence of the mutation; whereas, if the new mutation started with the more familiar mutation rate of 1 in 1,000,000 the whole business would be settled, with a considerable margin to spare, in the first generation. It is to be presumed that mutation rates, like the other characteristics of organisms, change only gradually in the course of evolution; whereas, however, the mutation rate of an unfavourable mutation will be allowed to increase up to 1 in 1,000,000 or even higher, without appreciably affecting the character of the species, favourable mutations can scarcely be permitted to continue occurring for long, even at rates 1,000-fold less, and we cannot exclude the possibility that a proportion of the favourable mutations that occur and are ultimately adopted, may have mutation rates so low that they occur sporadically, perhaps once only in thousands of generations. A quantitative comparison of the mutation rates current in homologous mutations in different allied species might well throw light on the difficult question as to how rapidly mutation rates should be thought of as increasing or decreasing.

When there is no selective advantage or disadvantage, the fraction of cases in which extinction has not taken place after n generations is, as has been seen, approximately $2/n$. It follows, since in the absence of selection the expectation in any future generation is equal to the number now living, that the average number of individuals in which these surviving genes will each be represented, is $\frac{1}{2}n$. This number will, however, vary greatly in different cases and it is of some interest to obtain the actual form of its distribution.

This can be done by observing that, if the frequency with which each number occurs is the coefficient of the corresponding power in the expansion of

$$\phi(x),$$

then substituting e^t for x, we have in ϕ the generating function of the moments of the distribution. Now to advance one generation is to substitute

$$e^{x-1} \text{ for } x$$
or
$$e^{e^t-1} \text{ for } e^t$$
or
$$e^t-1 \text{ for } t;$$

if therefore the frequency with which the number of individuals is k is represented by

$$\frac{2}{n^2}u(k/n)$$

where u is a continuous frequency function which tends, as n is increased, to a determinate limiting form, then

$$\phi(e^t) = \sum_{k=1}^{\infty} \frac{2}{n^2}u(k/n)e^{kt}$$

may be equated to

$$\frac{2}{n}M(nt)$$

where M is the characteristic function of the distribution u. If, when n is large, the form of this characteristic function is unchanged by random breeding for one generation, while the proportion of mutations not yet extinct falls from $2/n$ to $2/(n+1)$, then we may infer that M will satisfy the functional equation

$$\frac{2}{n}M\{n(e^t-1)\} = \frac{2}{n+1}M\{(n+1)t\} + \frac{2}{n(n+1)}.$$

Since n is to be increased without limit, let

$$e^t = 1 + \frac{\tau}{n}$$

so that

$$n(e^t-1) = \tau$$

$$(n+1)t = \tau + \frac{1}{n}\tau(1 - \tfrac{1}{2}\tau).$$

Neglecting terms in n^{-2}, then the equation for M reduces to the differential equation

$$M - 1 = \tau(1 - \tfrac{1}{2}\tau)\frac{dM}{d\tau}$$

or

$$\frac{dM}{M-1} = \frac{d\tau}{(1 - \tfrac{1}{2}\tau)} = \frac{d\tau}{\tau} + \frac{\tfrac{1}{2}d\tau}{1 - \tfrac{1}{2}\tau}$$

of which the integral is

$$\log(M - 1) = \log\tau - \log(1 - \tfrac{1}{2}\tau) + \text{constant}$$

or

$$M - 1 = \frac{c\tau}{1 - \tfrac{1}{2}\tau} \cdot$$

Since the mean value of k is $\tfrac{1}{2}n$, the mean of the distribution u must be $\tfrac{1}{2}$, and since this will be the coefficient of τ in the expansion of M, we find that $c = \tfrac{1}{2}$, and that

$$M = \frac{1}{1 - \tfrac{1}{2}\tau} \cdot$$

Knowing the characteristic function we may at once infer that

$$u(x) = 2e^{-2x}$$

and thence that

$$u\left(\frac{k}{n}\right) = 2e^{-2k/n}$$

and that the probability of exceeding k individuals is

$$\frac{2}{n}e^{-2k/n} \cdot$$

An inference of some interest is that in the absence of favourable selection, the number of individuals having a gene derived from a single mutation cannot greatly exceed the number of generations since its occurrence. Actually, the chance is less than 1 in 1,000 that x should exceed $3\tfrac{1}{2}n$. If, therefore, a mutant form exists in as many as 1,000 million individuals in each generation, we may be confident either that its numbers have been increased, at least up to a certain point, by selection, which is a relatively rapid process, or by recurrent mutation unopposed by selection, which must usually be a much slower process, or if we must suppose that it has originated in a single act of mutation and owes its present numbers to chance increases, that the process has been going on for at least 280 million generations, which makes it much the slowest and, for such high numbers, the least probable process of all.

A similar investigation of the distribution of the numbers, attained by the descendants of individual genes enjoying a small selective advantage, shows that the ultimate form of the distribution is the same in this case also. The probability of exceeding the number x after n generations may now be written

$$e^{-2(c-1)xc^{-n}}$$

showing of course that, as c exceeds unity, the numbers are certain to exceed any specified value of x in a sufficiently great number of generations.

The formula should represent the distribution correctly so long as

$$\frac{c^n}{2(c-1)}$$

is still a small fraction of the number of individuals in the species, but it evidently represents only the distribution of the numbers derived from mutations all of which occur in the same generation. This is an artificial and unnecessary limitation, since, as we have seen, with advantageous mutations those which occur earliest will first have an opportunity of establishing themselves, and will, after comparatively few trials, preclude the necessity for further mutations of the same sort. We must suppose that when favourable mutations occur they have seldom occurred before, and that their mutation rate is generally increasing. As to the nature of such increase we have no direct knowledge, but if it is dependent upon a change in the genotypic constitution of the species we must suppose it to be gradual, and since negative mutation rates are meaningless the simplest possible assumption is that the relative rate of increase per generation may be represented by a small number k, so that the mutation rate increases by the factor e^k in each generation.

On this assumption the number of mutations which at any stage are already represented in more than x individuals, will be proportional to

$$\int_0^\infty e^{-kn}\, e^{-2(c-1)xc^{-n}}\, dn$$

which turns out, when x is sufficiently large for $(c-1)x$ to be as great as 4 or 5, and large compared to $k/\log c - 1$, to be very nearly proportional to

$$x^{-k/\log c}.$$

In this formula we may recognize the element $\log c$, which is the amount by which the mutant gene avails to increase the Malthusian parameter of Chapter II, or approximately the selective advantage, 0·01, of our numerical illustrations. It measures the relative rate of increase of frequency of the gene in question, just as k measures that of its mutation rate. If it be supposed that the mutation rate depends wholly upon the presence of certain groups of genotypes, we must suppose k and $\log c$ to be quantities of the same kind, and of the same order of magnitude, but not necessarily approximately equal. If we consider that, of the gene substitutions capable of influencing any particular mutation rate, some may be progressing in one direction and some in the other, and that in general the increases in mutation frequency due to the increasing frequency of some genotypes, will be partly compensated by the disappearance of other genotypes in which the mutation also occurs, it appears probable that k must very frequently be the smaller quantity. If we confine attention to mutations possessing a selective advantage of just 1 per cent., this amounts to saying that when such mutations just begin to occur, the mutation rate is not increasing so rapidly as to double or treble itself within 100 generations, while not excluding the possibility that the increase in this period should be 10 per cent. or so. The practical consequence which follows if the ratio $k/\log c$ is small is that, of the mutant genes which ultimately pervade the species a large proportion are derived from that one individual mutation which first has the good fortune to establish itself in appreciable numbers, while only a negligible fraction can be contributed by the aggregate of all similar mutations which achieve a less or later success. Whereas if $k/\log c$ were large the mutant genes would be derived, though in unequal numbers, from a large number of separate mutations, no one of which would contribute a large fraction of the total.

It should be noticed that in respect to the initial stages in which survival is determined, c is the absolute rate of multiplication of the mutant type, and only approximately to be equated to its selective advantage over other genotypes. The difference becomes plain if we consider not, as hitherto, a stationary population, but one increasing or decreasing in numbers. In an increasing population mutations possessing no selective advantage, or indeed mutations at a selective disadvantage, provided this is less than the rate of increase of the species as a whole, will have a finite chance of avoiding extinction; while with a declining population, even mutations possessing a slight selective advantage, if this is less than the rate of decrease of the species, will be in a worse position than neutral mutations in a species of stationary size. In consequence growing populations receive greater accessions to their variability than stationary populations, while declining populations receive less; and if the intensity of selective actions is the same in both cases, we may expect growing populations to grow more variable, and declining populations to become less so by a process which is distinct from the effect of population size itself upon variability. In part at least the effect of increase will anticipate the consequences of the effect of size, for it will be shown that with larger populations statistical equilibrium will be established with a larger variance, and the direct effect of increasing population will be to increase the variance without waiting for the slower process of the establishment of a statistical equilibrium to show its effects.

The scope of this cause is limited by the actual rates of increase or decrease of natural populations, and I suppose that such changes are seldom so great as an increase of one-hundredfold in 10,000 generations, or about 1 in 2,000 in each generation over such a period. How important may be the contribution of mutations conferring an advantage or disadvantage of less than 1 in 2,000 is quite uncertain. It must certainly be greatest where adaptation, in the sense developed in Chapter II, is most intense, and it would at least be premature to assume that such minute changes are generally either rare, or without substantial evolutionary effects, although such may in fact be the case.

The distribution of gene ratio in factors contributing to the variance

We are now in a position to consider the relationships which must exist between the genetic variability maintained in a species, and the frequency of occurrence of mutations. The fundamental theorem proved in Chapter II will have prepared us to find that the variance maintained in fitness to survive must be intimately connected with the frequency of occurrence of favourable mutations; although a portion of it is generated by the occurrence of persistent unfavourable mutations of the kind considered in Chapter III, and is effective only in continually freeing the species from these defects. Such persistent unfavourable mutations will also contribute to the variance maintained in all other measurable characters, and further contributions must be supplied by those cases in which the gene ratio is in stable equilibrium under selective influences, to be considered more fully in Chapter V, and by cases in which the advantages of a character in one region or station occupied by the species are counteracted by disadvantages in alternative situations, a case the evolutionary consequences of which will be considered in Chapter VI. Our immediate purpose is to discuss the maintenance by mutations of that more elusive and fluid portion of the variance which is maintained by favourable mutations, and by those having a selective advantage or disadvantage so small that it may be neglected. Part of our problem will be to determine how small such selective advantage or disadvantage must be. The favourable mutations must, as was shown in Chapter II, be generally exceedingly minute in their somatic effects, and as we have seen in this chapter they must individually possess mutation rates so low that we are in fact confronted not with a calculable stream of mutations of each type, but with individual and sporadic occurrences. Mutations having nearly neutral effect might on the contrary have time to attain considerable mutation rates, for even if the rates were high, some million generations or more would be required to establish the new type, and this would give time for the mutation rate to rise from its initial inappreciable value. Apart from this slight difference the two cases may be treated together.

To distinguish the parts played by the different elements of the problem we need only consider three cases. First the distribution of gene ratio when, in the absence of selection or mutation, the variance is gradually decaying through the random extinction of genes. Next, the distribution when the variance is maintained by new mutations uninfluenced by selection; and finally the distributions appropriate to slight selective advantage or disadvantage.

The most powerful method of treating the first two of these problems is that of obtaining a functional equation for the series of terminal frequencies. If the number of individuals breeding in each generation is n, a large number of many millions or thousands of millions, the possible values of the gene frequency p are $1/2n$, $2/2n$, $\ldots$; these possible values are very numerous, and in the greater part of its range of distribution we may conveniently consider p as a continuous variate. At the extremes, however, a more exact treatment will be necessary, and here we shall make the simplifying assumption that the form of the terminal distribution, when statistical equilibrium is established, is not affected by the size of the population.

If now b_1, b_2, b_3, $\ldots$ stand for the frequencies at the values $p = 1/2n$, $2/2n$, $3/2n$, $\ldots$ we may define a function

$$\phi(x) = b_1 x + b_2 x^2 + b_3 x^3 + \ldots,$$

and the conditions of statistical equilibrium will yield a functional equation, the solution of which will give the frequencies b_1, b_2, b_3, $\ldots$, and therefore the distribution of the gene ratio. In the case of extinction without mutation, we may, in particular, ask what values the coefficients b must have in order that just one gene shall be exterminated in each generation. The sum of the values of these coefficients will then give the number of factors contributing to the variance, and from this we can determine the relation between the variance and its rate of decrease by random extinction.

If extermination takes place at the rate of one gene in each generation, we may suppose that half of these consist of cases in which the number of genes present is reduced from $1, 2, 3, \ldots$ to 0, and half to cases in which it is increased from $2n - 1$, $2n - 2$, $2n - 3$, $\ldots$ to $2n$. Genes represented in 0 individuals will of course supply the coefficient of x^0 in ϕ, so that after one generation the function ϕ representing the distribution at one terminal must be increased by $\frac{1}{2}$.

But in one generation we have already seen that $\phi(x)$ will be replaced by

$$\phi(e^{x-1});$$

consequently the equation to be satisfied by ϕ is

$$\phi(e^{x-1}) - \phi(x) = \tfrac{1}{2}.$$

To facilitate the solution of functional equations of this sort, it is necessary to consider a function u_ν of an argument ν such that

$$u_{\nu+1} = e^{u_\nu - 1}.$$

If this equation is satisfied by any function $f(\nu)$, it will evidently also be satisfied by $F(\nu) = f(\nu + k)$, consequently we may assign arbitrarily the value $u_0 = 0$, from which u_ν, if ν is any positive integer, may be obtained by direct substitution. In practice values of u for non integral ν are obtained by interpolating in the series of integral values, at about $\nu = 20$, and calculating lower values from the interpolates by means of the relation

$$u_{\nu-1} = 1 + \log u_\nu.$$

We may now write the functional equation for ϕ in the form

$$\phi\,(u_{\nu+1}) - \phi\,(u_\nu) = \tfrac{1}{2},$$

from which it appears that ϕ must be the same function of x as $\frac{1}{2}\nu$ is of u. The initial frequencies will therefore be obtained from the differential coefficients of ν with respect to u at $u = 0$, while the law of frequencies for larger values of p will be inferred from the behaviour of the function ν as u tends to unity.

Now, putting

$$v_\nu = \frac{1}{1 - u_\nu},$$

we have the recurrence formula

$$v_{\nu+1} = \frac{1}{1 - e^{-1/v_\nu}} = v_\nu + \tfrac{1}{2} + \frac{1}{12 v_\nu} - \frac{1}{720\, v^3} \cdots$$

so that as ν tends to infinity, u must tend to unity, and v_ν to

$$\frac{1}{2}\,\nu + \frac{1}{6}\,\log \nu + c$$

where the numerical value of c is found to be about $0 \cdot 899144$. The result shows that $\frac{1}{2}\nu(1 - u)$ tends to unity with u, and therefore that the frequency at $p = r/2n$ tends to unity as r is increased.

Moreover it follows that

$$\frac{1}{2}\,\nu = v - \frac{1}{6}\,\log v - c'$$

where c' tends to about $1 \cdot 014649$ as ν tends to infinity, consequently

$$\tfrac{1}{2}\nu - \frac{u}{1 - u} - \frac{1}{6}\,\log\,(1 - u)$$

tends to about $-0 \cdot 014649$ when $u = 1$. Apart from this finite portion of the frequency, the distribution is therefore given by the expansion in a Maclaurin series of

$$\frac{x}{1 - x} + \frac{1}{6}\,\log\,(1 - x)$$

or,

$$\frac{5}{6}\,x + \frac{11}{12}\,x^2 + \frac{17}{18}\,x^3 + \cdots,$$

and the coefficients of this series may be taken as a second approximation to the frequencies of factors, the rarer genes in which appear in 1, 2, 3, ... loci.

The actual values of the earlier coefficients may be obtained, though with decreasing precision by tabulating the function u_ν; these are shown in Table 3.

TABLE 3. *Terminal frequencies of factors suffering extinction.*

	Second Approximation.	Actual frequency.	Difference.
1 . .	0·833333	0·818203	−0·015131
2 . .	0·916667	0·916762	+0·000096
3 . .	0·944444	0·944923	+0·000479
4 . .	0·958333	0·958266	−0·000067
5 . .	0·966667	0·966634	−0·000033
6 . .	0·972222	0·972225	+0·000003

from which it appears that nearly the whole of the small discrepancy 0·014649 is accounted for by the first few terms, and that thereafter the frequency is well represented by the values $1 - 1/6r$. The terminal frequencies are shown in Fig. 6.

The total number of factors in such a distribution may now be estimated to be

$$2 \left\{ n - \frac{1}{6} (\gamma + \log 2n) - 0 \cdot 014649 \right\}$$

where γ is Euler's constant 0·577216. The remainder of this expression may be neglected in comparison with $2n$, so that the solution attained shows a decay of variance of only one part in $2n$ in each generation.

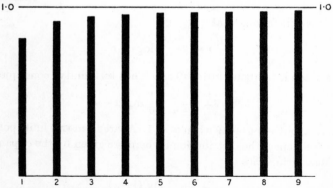

FIG. 6. Frequencies with which factors are represented by 1, 2, 3, . . . genes in the whole population, in the case of steady extinction without mutation. The upper line represents unit frequency at each value, which is approached for the higher values. Random survival will exterminate genes at the rate of one in every two generations, while leaving the distribution exhibited unchanged.

This is an extremely slow rate of decay; if the variance of species could be imagined to be ascribable to factors unaffected by selection, and if no new mutations occurred, the variance would decay exponentially so as to be reduced after τ generations in the ratio

$$e^{-\tau/2n}$$

it would therefore halve its value in $2n \log 2$, or about $1 \cdot 4n$ generations. No result could bring out more forcibly the contrast between the conservation of the variance in particulate inheritance, and its dissipation in inheritance conforming to the blending theory.

In a previous attack on this problem I was led by an erroneous method to the correct distribution for the factors contributing to the variance in a state of steady decay, but gave the time of relaxation as $4n$ instead of $2n$ generations. Professor Sewall Wright of Chicago, who had arrived by an independent method at the correct result, drew my attention to the discrepancy and has thus led me to a more exact examination of the whole problem.

The extremely slow rate of the natural decay of the variance is due to the fact that the great majority of factors possess gene ratios which are not extremely unequal. The distribution of z for this case is shown in Fig. 7, where it will be seen that for nearly all factors z lies between ± 6, and therefore that the rarer genes scarcely ever occupy less than $1/400$ of the loci available, and thus are in little danger of extinction.

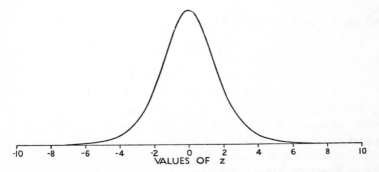

Fig. 7. Distribution of the measure of gene-ratio z, when the variance is in a state of steady decay, with neither mutations nor selection. The time of relaxation is now twice as many generations as the number of parents in each generation.

The method first developed has certain advantages for examining the frequency distribution in the central region. If θ is any measure of gene frequency, the frequency in any differential element $d\theta$, may be represented by $y\,d\theta$, and the condition of statistical equilibrium may be put in the form of a differential equation for the unknown function y. Using the variate defined by

$$\cos\theta = 1 - 2p$$

where θ is an angle in radian measure, which increases from 0 to π as p increases from 0 to 1, I obtained in 1922 the equation, like that of the conduction of heat,

$$\frac{\partial y}{\partial \tau} = \frac{1}{4n}\frac{\partial^2 y}{\partial \theta^2}$$

which with the solution $y = \sin\theta$, leads to a condition of steady decay with time of relaxation equal to $4n$ generations. The correct differential equation is, however,

$$\frac{\partial y}{\partial \tau} = \frac{1}{4n}\frac{\partial}{\partial \theta}(y\cot\theta) + \frac{1}{4n}\frac{\partial^2 y}{\partial \theta^2},$$

which while admitting the same solution yields the correct time of relaxation.

In the second case to be considered, in which the variance maintained is in statistical equilibrium with a constant supply of fresh mutations, we may apply this method at once by putting $\partial y/\partial\tau = 0$. Integrating the right hand side we obtain

$$\frac{\partial y}{\partial \theta} + y\cot\theta = A,$$

where A is some constant, whence

$$\frac{\partial}{\partial \theta}(y\sin\theta) = A\sin\theta,$$

$$y\sin\theta = -A\cos\theta + B,$$

$$y = B\operatorname{cosec}\theta - A\cot\theta.$$

The symmetrical solution makes y proportional to cosec θ. In the variate z this is a flat-topped distribution, all equal intervals dz being equally probable, at least in the central portion for which alone the differential equation is valid. Since when $\theta = \pi$, cosec θ + cot θ = 0 we may consider also the solution

$$y = B(\operatorname{cosec}\theta + \cot\theta)$$

appropriate to the case in which all mutations are taken to occur at $\theta = 0$.

In either case the integral over the whole range is infinite, owing to the rapid increase of y at $\theta = 0$. It does not follow that the total number of factors is infinite, for it is exactly in this region that the differential equation is invalid. In terms of p the frequency element (cosec $\theta + \cot \theta$) is equivalent to

$$\frac{2qdp}{2pq} = \frac{dp}{p}$$

so that the unsymmetrical solution obtained is one in which the frequency at $p = r/2n$ is proportional to $1/r$, at least when r is large. The total frequency will then evidently involve log $(2n)$, but to determine its value the examination of the terminal conditions is in this case also essential.

If $\phi(x)$ again represent the function, the coefficients of the expansion of which in powers of x are the frequencies maintained at $p = 1/2n$, $2/2n$, . . , by a single mutation in each generation, the functional equation for ϕ is now

$$\phi(e^{x-1}) - \phi(x) = 1 - x,$$

in which equation the left hand side represents the change in $\phi(x)$ due to random reproduction for one generation, while the effect of a single mutation must be to increase the coefficient of x by unity, and to reduce the absolute term (x^0) by unity. To solve the equation we may again utilize the device of writing u_ν for x, and obtain the equation

$$\phi(u_{\nu+1}) - \phi(u_\nu) = 1 - u_\nu.$$

Now, from the equation

$$u_{\nu+1} = e^{u_\nu - 1},$$

it appears on differentiating with respect to ν, that

$$u'_{\nu+1} = e^{u_\nu - 1} u'_\nu$$

or that $\log u'_{\nu+1} - \log u'_\nu = u_\nu - 1.$

Hence the equation for ϕ, may be written

$$\phi(u_{\nu+1}) - \phi(u_\nu) = -(\log u'_{\nu+1} - \log u'_\nu)$$

an equation which is satisfied if $\phi(u_\nu)$ differs from $-\log u'_\nu$ by a constant. The constant part of $\phi(x)$, representing the frequency of the factors not represented in any individual is of course arbitrary, and on the convention that $\phi(0) = 0$, we have the solution

$$\phi(u_\nu) = \log u'_0 - \log u'_\nu$$

or, if ν' stands for the differential coefficient of ν with respect to u

$$\phi(u) = \log \nu' - \log \nu'_0$$
$$= \log \nu' - 0 \cdot 492502$$

this being an empirical evaluation of the constant term.

Now as u approaches unity, we have seen that ν increases proportionately to $2/(1-u)$, and therefore $\log \nu'$ tends to equality with $\log 2 - 2 \log (1-u)$; apart from a finite discrepancy in the terminal frequencies, the frequencies will be given by the coefficients of the expansion

$$- 2 \log (1-x) = 2x + \frac{2}{2} x^2 + \frac{2}{3} x^3 + \ldots,$$

so that the frequency at $p = r/2n$ approaches $2/r$ as r is increased, in accordance with the solution found from the differential equation.

The first few actual coefficients are:

TABLE 4.

Terminal frequencies for factors maintained by mutations.

	Approximation.	Actual.	Excess.
1	2·000000	2·240917	+0·240917
2	1·000000	0·953776	−0·046224
3	0·666667	0·671864	+0·005197
4	0·500000	0·501096	+0·001096
5	0·400000	0·399762	−0·000238

The total number of factors maintained in the population by one new mutation in each generation will be the sum of the $2n$ first coefficients of the expansion of $\phi(x)$, or

$$2(\gamma + \log 2n) + 0·200645.$$

For values of n from a million to a billion, the following table shows

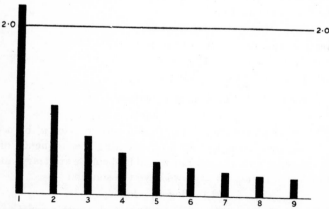

FIG. 8. Frequencies with which factors are represented by 1, 2, 3, . . . genes in the whole population, in the case when the variation is maintained by fresh mutations at a constant level. For one new mutation in each generation the frequency for r genes is nearly $2/r$.

the number of factors contributing to the specific variance for each one occurring per generation:

TABLE 5.

n.	Number of factors.
10^6	30·4
10^7	35·0
10^8	38·6
10^9	44·2
10^{10}	48·8
10^{11}	53·4
10^{12}	58·0

Fig. 8 shows the distribution of the terminal frequencies. It will be observed that a considerable 'head' of new mutations is needed to maintain even low frequencies at the central values. The number of these central values is, however, so great that the numbers maintained even by only a single mutation in each generation are, as table 5 shows, considerable, and practically proportionate to the range in the values of z possible for a population of given size.

In the third case to be considered, the frequency of a gene is favoured by selection so that $\log p/q$ increased in each generation by an amount a supposed small, then in one generation

$$\delta p = apq$$

and

$$\delta \theta = a\sqrt{pq} = \tfrac{1}{2} a \sin \theta.$$

The effect of selection is thus to produce a flux $\tfrac{1}{2} ay \sin \theta$, and our differential equation takes the form

$$\frac{\partial y}{\partial \tau} = -\tfrac{1}{2} a \frac{\partial}{\partial \theta} (y \sin \theta) + \frac{1}{4n} \frac{\partial}{\partial \theta} (y \cot \theta) + \frac{1}{4n} \frac{\partial^2 y}{\partial \theta^2}.$$

For statistical equilibrium, maintained by mutations, we now require that

$$\frac{\partial y}{\partial \theta} + y \cot \theta - 2 \, any \sin \theta = A$$

which may be put in the form

$$\frac{\partial}{\partial \theta} (y \sin \theta e^{2 \, an \cos \theta}) = A \sin \theta e^{2 \, an \cos \theta}$$

or

$$y \sin \theta e^{2 \, an \cos \theta} = \int A \sin \theta e^{2 \, an \cos \theta} + B;$$

performing the integration, this leads to

$$y = \operatorname{cosec} \theta \left(\frac{A}{2an} + Be^{-2 \, an \cos \theta} \right)$$

The value of the flux

$$\frac{1}{2}\, ay\sin\theta - \frac{1}{4n}\, y\cot\theta - \frac{1}{4n}\frac{\partial y}{\partial\theta} = \frac{A}{4n}$$

is $\frac{1}{2}a$ times the coefficient of cosec θ.

The solution appropriate to a supply of mutations at the rate of one in each generation having each a small selective advantage a, must be equal to 4 cosec θ at $\theta = 0$, while at $\theta = \pi$ where no mutations are occurring, it must be proportional to sin θ. The appropriate form is

$$y = \frac{4\cosec\theta}{1 - e^{-4\,an}}\left\{1 - e^{-2\,an\,(1+\cos\theta)}\right\},$$

for which the frequency in the range dp is

$$\frac{2\,dp}{pq}\frac{1 - e^{-4\,anq}}{1 - e^{-4\,an}}\cdot$$

When $q = 1$ this evidently gives a terminal distribution similar to that given by mutations without selective advantage, but maintained by mutation, while q tends to zero, we have

$$\frac{8\,an}{1 - e^{-4\,an}}\,dp,$$

appropriate to extinction without mutation, the rate of extinction being $2\,a/(1 - e^{-4\,an})$, which must represent the probability of ultimate success to a mutation with small selective advantage a. When $a = 0$, this probability tends to the limiting value $1/2\,n$, which is the probability of success of a mutation without selective advantage, and is not effectively increased so long as $4\,an$ is a small quantity; if $4\,an$ is neither large nor small the full formula is required, but if $4\,an$ is large, the exponential factor is negligible, and the probability of success is given very nearly by $2\,a$. The frequency distribution in this case is represented by

$$\frac{2\,dp}{pq}\,(1 - e^{-4\,anq})$$

in which the second term is only appreciable for small values of q, where a constant frequency $8an\,dp$, or $4a$ in each possible value of p is maintained, as is appropriate to the extinction of $2a$ in each generation.

The differential equation is still valid for values of a which make an large, but requires that a^2n should be small. The exact treatment of selective rates less extremely small than those here dealt with would not probably differ essentially from that appropriate to very small selections.

The corresponding solution for the equally important case of mutations with a small selective disadvantage, may be found by changing the sign of a. The chance of success is now always less than $1/2n$, being

$$\frac{2a}{e^{4an} - 1} ,$$

and is in all cases negligible. The distribution has a frequency in the range dp

$$\frac{2dp}{pq} \frac{e^{4anq} - 1}{e^{4an} - 1}$$

which, when an is large, is simply

$$\frac{2dp}{pq} e^{-4anv}$$

giving a total number of factors nearly $2 \log (1/2a)$, so long as $4an$ is large, for each such mutant per generation.

Thus disadvantageous mutations, unlike those which are advantageous, or practically neutral (an small) maintain no more factors contributing to the variance of numerous species than of rare species. The calculations refer, of course, to the variation maintained by a fixed number of mutations, and take no account of the fact that in abundant species there will be many more individuals in which mutations may occur, in each generation.

The analysis shows how very minute must be the selective intensity acting on a factor, before we can count it as neutral either for the purpose of evaluating the probability of a sporadic mutation establishing itself in a species, or in considering the relation between mutation rate and the variance maintained. In either case we are concerned with the product an found by multiplying the selective advantage by the number breeding in each generation. In respect of survival small deviations of this quantity from zero exert a considerable effect, the chance of survival, for example, is increased more than fiftyfold as an increases from -1 to $+1$. The contribution to the variance is less sensitive, since this depends little on the terminal frequencies and principally on the central frequencies; broadly speaking neutral mutations contribute half as much to maintaining the variance as is contributed by those with a substantial selective advantage. If an is $-2 \cdot 5$ the contribution is a tenth, while at $+2 \cdot 5$ it is nine-tenths of the full value. Evidently in a population of a thousand million, only those gene contrasts, which possess an equipoise of advantage within at most a few parts in a thousand million, can be regarded as neutral. The distribution for favourable mutations from these minute advantages up to advantages a millionfold greater must be all very similar, the probability of falling in equal ranges dz being nearly constant over the whole range of possible values. Disadvantageous mutations are confined to smaller and smaller values of the gene ratio as the disadvantage increases, until while the disadvantage is still very minute, the only appreciable contribution will be that made by mutations having appreciable or high mutation rates.

Although for the same number of neutral or beneficial mutations per generation, abundant species will maintain a larger number of factors contributing to the variance, than will rarer species, yet this is due principally to the greater range of the values of z available. The additional factors will then have somewhat extreme gene ratios, and will therefore contribute little to the measurable variance. The great contrast between abundant and rare species lies in the number of individuals available in each generation as possible mutants. The actual number of mutations in each generation must therefore be proportional to the population of the species. With mutations having appreciable mutation rates, this makes no difference, for these will reach an equilibrium with counterselection at the same proportional incidence. The importance of the contrast lies with the extremely rare mutations, in which the number of new mutations occurring must increase proportionately to the number of individuals available. It is to this class, as has been shown, that the beneficial mutations must be confined, and the advantage of the more abundant species in this respect is especially conspicuous.

The very small range of selective intensity in which a factor may be regarded as effectively neutral suggests that such a condition must in general be extremely transient. The slow changes which must always be in progress, altering the genetic constitution and environmental conditions of each species, must also alter the selective advantage of each gene contrast. Slow as such changes in selective advantage must undoubtedly be, the zone separating genes possessing a definite selective advantage from those suffering a definite selective disadvantage is so narrow, of the order of the reciprocal of the breeding population, that it must be crossed somewhat rapidly. Each successful gene which spreads through the species, must in some measure alter the selective advantage or disadvantage of many other genes. It will thus affect the rates at which these other genes are increasing or decreasing, and so the rate of change of its own selective advantage. The general statistical consequence is that any gene which increases in numbers, whether this increase is due to a selective advantage, an increased mutation rate, or to any other cause, such as a succession of favourable seasons, will so react upon the genetic constitution of the species, as to accelerate its increase of selective advantage if this is increasing, or to retard its decrease if it is decreasing. To put the matter in another way, each gene is constantly tending to create genetic situations favourable to its own survival, so that an increase in numbers due to any cause will in its turn react favourably upon the selective advantage which it enjoys.

It is perhaps worth while at this point to consider the immense diversity of the genetic variability available in a species which segregates even for only 100 different factors. The total number of true-breeding genotypes into which these can be combined is 2^{100}, which would require 31 figures in the decimal notation. The number including heterozygotes would require 48 figures. A population of a thousand million or a billion individuals can thus only exhibit the most insignificant fraction of the possible combinations, even if no two individuals are genetically alike. Although the combinations which occur are in all only a minute fraction of those which might with equal probability have occurred, and which may occur, for example, in the next generation, there is beyond these a great unexplored region of combinations none of which can be expected to occur unless the system of gene ratios is continuously modified in the right direction. There are, moreover, millions of different directions in which such modification may take place, so that without the occurrence of further mutations all ordinary species must already possess within themselves the potentialities of the most varied evolutionary modifications. It has often been remarked, and truly, that without mutation evolutionary progress, whatever direction it may take, will ultimately come to a standstill for lack of further possible improvements. It has not so often been realized how very far most existing species must be from such a state of stagnation, or how easily with no more than one hundred factors a species may be modified to a condition considerably outside the range of its previous variation, and this in a large number of different characteristics.

Self-sterility Allelomorphs

A somewhat peculiar problem is offered by plants having a series of self-sterility allelomorphs so constituted that pollen is acceptable only if it contains an allele different from either of those in the seed parent. This case well illustrates the aptitude for such problems of the method of differential equations, with proper attention to the interpretation of the terminal conditions.

If the frequency of any allele i is p_i, so that taking all alleles into account

$$S_i(p_i) = 1$$

and if the frequency of the genotype involving alleles i and j is p_{ij}, so that

$$S_j(p_{ij}) = 2p_i$$

then the frequency to be expected for all genotypes in a population derived by random crossing, from one in which the frequencies of all genotypes are known, may be derived from the consideration that the proportion of ovules containing the gamete i exposed to pollination in plants (ik) will be $\frac{1}{2}p_{ik}$, and of these the fraction fertilized by j pollen is $p_j(1 - p_i - p_k)$, so that the number of (ij) plants expected which derive the allele i from their seed parent must be

$$(\tfrac{1}{2}p_j S_k) \frac{p_{ik}}{1 - p_i - p_k} \qquad\qquad k \neq i, j$$

Similarly the number which derive the allele j from the seed parent will be

$$(\tfrac{1}{2}p_i S_k) \frac{p_{jk}}{1 - p_j - p_k} \qquad\qquad k \neq i, j$$

and the sum of these two expressions gives the total proportion of genotype (ij) expected.

We have thus a recurrence relation expressing the frequencies of all genotypes expected in the next generation in terms of the frequencies existing at any stage. Unless the genotype frequencies can be expressed in terms of the gene frequencies p_i, we cannot, however, obtain corresponding recurrence relations for the gene frequencies.

The first expression, however, summed for all values of j other than i and k, gives

$$\tfrac{1}{2}S_k(p_{ik}) = p_i$$

illustrating the fact that the system exerts no selection through the ovules, but only through the pollen. The second expression similarly summed is

$$\tfrac{1}{2}S_{jk}(p_i)\frac{p_{jk}}{1 - p_j - p_k}$$

and this is not expressible in terms of the gene frequencies p_i.

. Although influenced by the frequencies of individual genotypes, the expression is not greatly influenced by any variation in these frequencies compatible with given gene frequencies, especially when all gene frequencies are small. Thus in comparing different genes i, a factor of major importance is the gene frequency p_i, and a second important factor is $S_{jk}(p_{jk})$, the total frequency of plants accessible to i pollen. These plants are indeed accessible in slightly varied degree, as is indicated by the factor $(1 - p_j - p_k)$, and consequently the accessibility of the aggregate of genotypes accessible to i pollen will differ, even more slightly, from those accessible to j pollen. However, a good approximation is supplied by ignoring these differences.

If then the contribution of gene i through pollen is taken to be proportional to

$$p_i S_{jk}(p_{jk}) = p_i(1 - 2p_i)$$

then the sum for all genes i is $(1 - 2\alpha)$, where

$$\alpha = S_i(p_i^2)$$

and the proportion of the next generation receiving gene i through the pollen is

$$\frac{p(1 - 2p)}{1 - 2\alpha}$$

Adding those receiving the same gene through the ovule, we have

$$p + \frac{p(1 - 2p)}{1 - 2\alpha} = \frac{2p(1 - p - \alpha)}{1 - 2\alpha}$$

so that the proportion of i genes in the new generation is

$$\frac{p(1 - p - \alpha)}{1 - 2\alpha}$$

which is less than the proportion in the parent generation by

$$p - \frac{p(1 - p - \alpha)}{1 - 2\alpha} = \frac{p(p - \alpha)}{1 - 2\alpha} .$$

The important point, which allows of a general solution, is that the change in gene ratio is approximately equal to $p(p - \alpha)$; a change in the denominator, small compared with α, or even of the same magnitude, in the expression for the change in gene ratio, will not appreciably affect any practical conclusions.

Selection thus acts adversely on all alleles with frequencies exceeding α but favourably on all alleles less frequent.

If α is a small quantity, the selective intensities may be treated as small, and the random sampling variance for plant number about the new expectation will be

$$V(2p) = 2p(1 - 2p)/N$$

Let now

$$\sin^2 \phi = 2p$$

$$\cos^2 \phi = 1 - 2p$$

$$\theta = 2\phi$$

and let

$$y \, d\theta = \frac{4y \, dp}{\sin \theta}$$

stand for a frequency element of alleles in the range $d\theta$.

The flux, evaluated as on page 92, is expressed by

$$\frac{1}{2N} \frac{\partial y}{\partial \theta} \frac{1}{2N} y \cot \theta + \frac{4p(p - \alpha)}{\sin \theta (1 - 2\alpha)} y$$

which for statistical equilibrium is to be equated to zero.

But

$$\int \cot \theta \, d\theta = \log \sin \theta$$

and

$$\int \frac{4p(p - \alpha)}{(1 - 2\alpha) \sin \theta} d\theta = \int \frac{16p(p - \alpha)}{(1 - 2\alpha) \sin^2 \theta} dp$$

$$\int \frac{2(p - \alpha)}{(1 - 2\alpha)(1 - 2p)} dp = \int \left\{ \frac{1}{1 - 2p} - \frac{1}{1 - 2\alpha} \right\} dp$$

Hence $(y \sin \theta)(e^{-2Np(1 - 2\alpha)})(1 - 2p)^{-N}$ is constant.

The element of frequency, therefore, $y \, d\theta$, may be written

$$\left(\frac{4B \, dp}{\sin^2 \theta} \right) (e^{-N(1 - 2p)/(1 - 2\alpha)}) (1 - 2p)^{+N}$$

$$= \left(\frac{B \, dp}{2p} \right) (e^{-N(1 - 2p)/(1 - 2\alpha)}) (1 - 2p)^{N - 1}$$

where B is a constant of integration, to be evaluated from the condition

$$S(p) = 1$$

As in the more general analysis, the different alleles can actually have only the discontinuous series of frequencies 1, 2, 3, 4 ... 2N, so that for very rare alleles, where the integral expression involving the continuous variable θ appears to diverge, we have in reality the finite series of frequencies represented by Fig. 8. The total effect of the deviations of these frequencies from a harmonic progression was there seen to be trifling. With an exact harmonic progression, corresponding to the factor $1/p$ in the distribution function obtained above, each member of the series would contribute an equal, but very small, fraction to the (unit) total frequency of all alleles. Taking into consideration the other factors of the distribution function the series is seen to be one diminishing exponentially as the terminus is approached. The exact contribution of these terminal frequencies must be closely equivalent to that of the tail of such a curve extended indefinitely. We shall therefore integrate with respect to $1 - 2p$ between the limits 0 and ∞.

Multiplying the element of frequency by p and integrating, it appears that

$$\tfrac{1}{4}B(N-1)!\left(\frac{1-2\alpha}{N}\right)^{N} = 1$$

and the absolute value of the frequency element is

$$\left(\frac{1}{p}\right)\left(\frac{2N\,dp}{1-2\alpha}\right)\left(\frac{N(1-2p)}{1-2\alpha}\right)^{N-1}\left(e^{-N(1-2p)/(1-2\alpha)}\right)\left(\frac{1}{(N-1)!}\right)$$

or

$$\left(\frac{1}{p}\right)\left(\frac{1}{(N-1)!}\right)x^{N-1}e^{-x}\,dx \qquad\qquad x>0$$

where x stands for

$$\frac{N(1-2p)}{1-2\alpha}$$

with the understanding that when p becomes small the continuous Eulerian distribution is replaced by the terminating series

$$p = \frac{1}{2N},\ \frac{1}{4N} \cdots \frac{1}{2rN} \cdots$$

where r is the number of representations of a particular allelomorphic gene. The constant of integration was obtained from the complete Eulerian distribution, and that it fulfills the analytical requirements of the problem may be shown by the evaluation of the sum of p^2 for all alleles. For

$$\int_0^\infty p\frac{1}{(N-1)!}x^{N-1}e^{-x}\,dx$$

when

$$p = \frac{1}{2}\left\{1 - \frac{(1-2\alpha)x}{N}\right\}$$

comes to

$$\frac{1}{2} - \frac{1-2\alpha}{2} = \alpha$$

in exact accordance with the manner in which α was originally defined.

At the terminus

$$p = \frac{1}{2N}, \qquad x = \frac{N-1}{1-2\alpha}, \qquad dx = \frac{2N}{1-2\alpha}dp$$

so the terminal frequency is

$$\left(\frac{1}{(N-1)!}\right)(N-1)^{N-1}(2N)(e^{-(N-1)/(1-2\alpha)})(1-2\alpha)^{-N}$$

but, to a close approximation,

$$(N-1)! = (N-1)^{N-1}e^{-(N-1)}\sqrt{2\pi(N-1)}$$

giving the terminal frequency as

$$\left(\frac{2N}{\sqrt{2\pi(N-1)}}\right)(e^{-2\alpha(N-1)/(1-2\alpha)})(1-2\alpha)^{-N}$$

which, to a sufficient approximation, may be simplified to

$$2\sqrt{\frac{N}{2\pi}}(e^{-2N\alpha^2})$$

the half of which must represent the number of new mutations required on the average in each generation.

The numerical results of this solution are simple. Consider the case of a population of only 10,000, with $\alpha = \cdot03$, a value corresponding with somewhat more than 35 alleles.

Then

$$2n\alpha^2 = 18$$

$$e^{-2N\alpha^2} = \cdot000000152$$

$$\sqrt{N/2\pi} = 39\cdot89$$

so that the loss of alleles to be expected by chance extinction is about 6·06 in a million generations.

This loss would be counterbalanced by a mutation rate to new or extinct alleles of about one in three thousand million; consequently, unless the mutation rate is exceedingly low, more alleles would accrue, and the value of α would fall below ·03.

In the case of *Oenothera organensis* the existing wild population has been thought to be so small as one thousand individuals, with about 37 alleles. In this case

$$2N\alpha^2 = 1\cdot8$$

$$e^{-2N\alpha^2} = \cdot165$$

$$\sqrt{N/2\pi} = 12\cdot622$$

and the estimated loss of alleles is as much as two per generation. A mutation rate of about one in a thousand would be needed to maintain in equilibrium the observed number of alleles in so small a population as this.

Species having populations less than 10,000 must, of course, be presumed to have fallen greatly in population during their recent evolutionary history, so it is not surprising that *O. organensis* should have α so high as ·03, high as this is.

Higher values of α than this seem not to have been observed in other species, and it may be confidently presumed that the value of α has risen as the population has fallen, and that it is not now even approximately in equilibrium with the mutation rate. Many of the rarer alleles have doubtless dropped out, leaving predominantly those well represented in the population, and consequently in little danger of immediate extinction.

Sewell Wright has proposed that the number of alleles observed might be the sum of the numbers of alleles in different highly isolated groups into which the totality of this small population is divided. This should be the case if the larger population has in its recent history become subdivided. Isolation, however, of the degree required, if it now exists, would necessarily be a recent condition due doubtless to the recent large reduction in population numbers.

It may be added that the treatment of this case by S. Wright[1] leads to results, as shown by his graphs, very different from those obtained above. Wright, however, fails to develop any explicit formulae, but seems to have relied on extensive numerical calculations based on trial values of numerous constants he introduces. It is hoped that the fore-

going discussion using the method of the 1930 edition will set the situation
of these alleles in a clearer light.

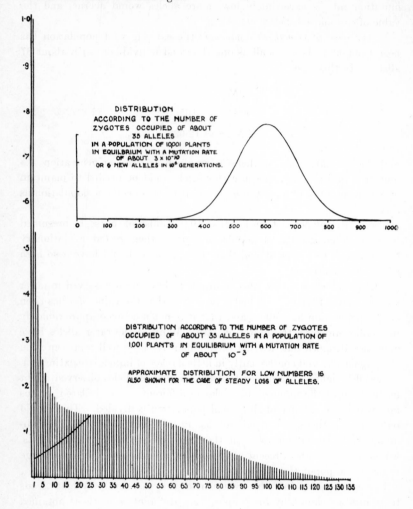

DISTRIBUTION
ACCORDING TO THE NUMBER OF
ZYGOTES OCCUPIED OF ABOUT
35 ALLELES
IN A POPULATION OF 1000I PLANTS
IN EQUILBRIUM WITH A MUTATION RATE
OF ABOUT 3 × 10⁻¹⁰
OR 6 NEW ALLELES IN 10⁶ GENERATIONS.

DISTRIBUTION ACCORDING TO THE NUMBER OF ZYGOTES
OCCUPIED OF ABOUT 35 ALLELES IN A POPULATION OF
1,001 PLANTS IN EQUILIBRIUM WITH A MUTATION RATE
OF ABOUT 10⁻³

APPROXIMATE DISTRIBUTION FOR LOW NUMBERS IS
ALSO SHOWN FOR THE CASE OF STEADY LOSS OF ALLELES.

Reference

1. S. Wright, 1939. 'The distribution of self-sterility alleles in populations.' *Genetics*,
    24, 538–52.

V

VARIATION AS DETERMINED BY MUTATION
AND SELECTION (continued)

The observed connexion between variability and abundance

In the second chapter of the *Origin of Species* Darwin summarizes a study of the causes of variability, based upon a statistical investigation of the number of well-marked varieties recorded in different species of plants. He was, perhaps unfortunately, dissuaded from publishing his actual tabulations, but gained the concurrence of Hooker to the general conclusions that 'Wide ranging, much diffused, and common species vary most'. Darwin was concerned to show that it was not merely that wide ranging forms give rise to local varieties in reaction to different inorganic and organic environments, but also that, 'In any limited country, the species which are most common, that is, abound most in individuals, and the species which are most widely diffused within their own country (and this is a different consideration from wide range, and to a certain extent from commonness), oftenest give rise to varieties sufficiently well marked to have been recorded in botanical works'.

A few years ago it was my privilege to make a statistical investigation of the extensive observations of Mr. E. B. Ford upon the variability of the wing colour in a number of species of night-flying moths. For thirty-five species the tints were sufficiently comparable to be represented on a single colour scale, and for these the observations, which included over 5,000 individuals, offered an exceptionally fine opportunity of examining the association between abundance and variability. It is essential in such an investigation to eliminate any tendency for one group of species to appear more variable than another owing to the peculiarities inherent in an arbitrary scale of tints. The data, however, were sufficiently copious to make it possible to eliminate this source of error, and after making the necessary allowances, it appeared that, in both sexes, the ten species classed as 'abundant' or 'very common' exceeded in variance the thirteen species which were less than common by between 70 and 80 per cent. the twelve 'common' species being in both cases of intermediate variability.

Because many other factors besides numbers must influence the variability, and particularly because the precision of any classification of abundance must be exceedingly low, it is essential to base such comparisons upon as large a number of species as is possible, and this seems to be an important cause of the present lack of satisfactory data bearing upon the variability of species. The differences observed among the moths were, however, sufficiently substantial to be statistically significant, even in comparison with the large differences in variability found within each class. It may be mentioned that the same data showed in fact a larger variability in the species with the widest geographical ranges, as contrasted with less widespread species, although the differences in this comparison cannot claim to be statistically established. There is no reason, however, to believe that an increase of numbers by increase of range is less effective in increasing variability than would be the same increase of numbers due to greater density of population, for the numerical ratio of species classed by entomologists as abundant and rare respectively must be much greater than is ordinarily the ratio of the areas occupied by different species.*

The theoretical deduction that the actual number of a species is an important factor in determining the amount of variance which it displays, thus seems to be justified by such observations as are at present available. Its principal consequence for evolutionary theory seems to be that already inferred by Darwin, that abundant species will, *ceteris paribus*, make the most rapid evolutionary progress, and will tend to supplant less abundant groups with which they come into competition. We may infer that in the ordinary condition of the earth's inhabitants a large number of less abundant species will be decreasing in numbers, while a smaller number of more abundant species will be increasing—the number of species being maintained by fission of the more abundant, and especially of the more widespread species, a subject which will be considered in the next chapter. It may be noted, however, that whereas an increase in fitness when invested in an increase in numbers in the manner described in Chapter II, is now seen to bear a substantial rate of interest in laying the foundations of sufficiently rapid further improvement, the process

*In 1937 also I was able to demonstrate an analogous relationship among about 180 species of British-nesting birds (Proc. Roy. Soc. B, 122, pp. 1-26).

of fission, while yielding doubtless an immediate adaptive advantage, yet entails a certain loss in the degree of variability which the divided parts can severally maintain. There is thus in the continuous elimination of the smaller specific groups a natural check set to the excessive comminution of species, which would ensue upon specialization in the direction of minutely differentiated aptitudes.

Among the factors which influence the relationship between variation and selection may be mentioned the tendency of like to mate with like, known as homogamy. Among the higher animals we have no certain knowledge of this save in the case of man, but this can scarcely detract from the value of the human evidence, since its occurrence is contrary to popular opinion, and not sufficiently explained by any circumstance of social organization. In collections of human measurements the resemblance between married persons is rather a conspicuous feature. Its principal biometric effects seem to be to increase the genetic variance produced by a given number of Mendelian factors with given gene ratios, and so to increase in a fixed proportion the intensity of the selection to which each is exposed. The effect of selection in human stature is increased in this way by more than 20 per cent. It is therefore potentially an important agent in promoting evolutionary change. Its causes are quite uncertain. It has been suggested that fertility depends in some measure upon the constitutional similarity of the mates, but evidence for this is lacking in the case of man and the higher animals, and I know of no serious attempt to demonstrate the truth or falsity of the suggestion. It is at least equally possible that the standards of sexual preference are slightly modified by individual size, and on this view the causes of homogamy can scarcely be distinguished from those which, in sexual preference, to be considered in Chapter VI, produce direct selective effects.

Stable gene ratios

We have hitherto considered only those factors which contribute to the genetic variance in fitness to survive and progress. There remain to be considered those factors in which one gene has a selective advantage only until a certain gene-ratio is established, while for higher ratios it is at a selective disadvantage. In such cases the gene ratio will be stable at the limiting value, for the selection in action will tend to restore it to this value whenever it happens to be disturbed from it in either direction. At this value the effect of the gene substitution upon survival will be zero, and consequently no contribution will be made to the genetic variance in fitness, although the genetic variance in other measurable characters may be augmented by such factors. These cases have a special importance owing to the principle that factors will be found most frequently when their rate of change in gene ratio is least. In consequence of this, if their stability could be assumed to be absolutely permanent, such cases would have been accumulating in each species since its earliest beginnings; in fact, however, the conditions of stability must themselves be transient during the course of evolutionary change, and we can only be sure that cases of such gene stability must exist with a frequency quite disproportionate to the probability of occurrence of the conditions on which the stability is based.

A single factor may be in stable equilibrium under selection if the heterozygote has a selective advantage over both homozygotes. For if we suppose the three phases of the factor to appear in any generation in the ratio $p^2 : 2pq : q^2$, and that their relative selective advantages are respectively in the ratio $a : b : c$, then the three phases in this generation will reproduce in the ratio $ap^2 : 2bpq : cq^2$, where the absolute magnitudes of the quantities a, b, c are a matter of indifference, only their ratios being required. If equilibrium in the gene ratio is established this ratio will be the same in those which reproduce as it was in the preceding generation, and therefore,

$$\frac{p}{q} = \frac{ap^2 + bpq}{bpq + cq^2};$$

whence it appears that

$$ap + bq = bp + cq.$$

Subtracting each of these from $b\,(p+q)$ we obtain

$$p\,(b-a) = q\,(b-c),$$

or

$$\frac{p}{q} = \frac{b-c}{b-a}.$$

There is therefore always a real ratio of equilibrium if $b-a$ and $b-c$ are either both positive or both negative; that is, if the heterozygote is either better or worse adapted than both the homozygotes. *A priori* we should judge either condition to be exceptional; they will not, however, be found in nature equally infrequently, for when

b is less than a and c the equilibrium is unstable and there will be no tendency for such cases to accumulate, whereas if b exceeds a and c the equilibrium is stable and such cases will therefore persist until the stability is upset.

To demonstrate the condition for stability it is sufficient to observe that the ratio

$$\frac{ap^2 + bpq}{bpq + cq^2}$$

may be written

$$\frac{p(ap + cq) + pq(b - c)}{q(ap + cq) + pq(b - a)}$$

which lies between the ratios $p : q$ and $(b - c) : (b - a)$, if b exceeds a and c, but not if $b - c$ and $b - a$ are negative.

In organisms capable both of self- and of cross-fertilization, the situation in which the heterozygote has a selective advantage tends to give the offspring by cross-fertilization a higher reproductive value than offspring by self-fertilization, and therefore to make it worth a somewhat greater expenditure; for in a population mating at random, such as is assumed above, the reproductive values of the three genotypes will be simply in the ratio $a : b : c$. Hence for the genotype of the first kind the average value of its offspring will be $pa + qb$, if it is cross-fertilized at random, against a if it is self-fertilized. For the heterozygote we find $\frac{1}{2}(pa + b + qc)$ for cross-fertilization, against $\frac{1}{4}(a + 2b + c)$ for self-fertilization. The average advantages in the two homozygous phases are thus $q(b - a)$ and $p(b - c)$ respectively, while in the heterozygote it is $\frac{1}{4}(p - q)(a - c)$. Remembering that the frequencies with which these three phases occur are in the ratio $p^2 : 2pq : q^2$ we find for the average loss of value in self-fertilization

$$\frac{1}{2}pq\{2p(b - a) + (p - q)(a - c) + 2q(b - c)\}$$
$$= \frac{1}{2}pq(2b - a - c).$$

Now according to our previous solution, equilibrium will be established when

$$p = \frac{b - c}{2b - a - c}, \quad q = \frac{b - a}{2b - a - c}$$

and if we are to interpret $a, b,$ and c as proportionate contributions to the ancestry of future generations we must have also

$$p^2a + 2pqb + q^2c = 1,$$

in which, if we substitute for p and q, we shall find the relation

$$b^2 - ac = 2b - a - c.$$

Using these relations, we may express the average loss of value of the offspring, caused by self-fertilization, as a homogeneous expression in a, b, and c only, in the form

$$\frac{(b-a)\,(b-c)}{2\,(b^2 - ac)}.$$

Thus, for example, if for any factor a, b, and c were in the ratio 5 : 6 : 4 a stable genetic situation would be established in which the products of self-fertilization would be worth, in respect of their prospects of contributing to future generations, just $\frac{1}{16}$ less than the average products of cross-fertilization. Any other factors of the same kind, which might happen to be present, would of course add to the advantage of cross-fertilization. The formula, however, given above, is that appropriate to organisms in which cross-fertilization is the rule, for if self-fertilization is much practised the reproductive values of the three phases will be in a higher ratio than their selective factors for a single generation.

Equilibrium involving two factors

Two factors, the alternative genes in which may be represented by A, a and B, b may maintain each other mutually in genetic equilibrium, if the selective advantage of A or a is reversed when B is substituted for b, or *vice versa*. Without attempting to specify the exact selective advantage enjoyed by each of the nine genotypes we may specify the type of selection under consideration by saying that A is advantageous in the presence of B but disadvantageous in the presence of b, and that B is advantageous in the presence of A but disadvantageous in the presence of a. Equally of course in this statement we might transpose the words advantageous and disadvantageous.

Equilibrium in such a system evidently implies that the increase in the frequency of A which takes place in the presence of B shall be exactly counterbalanced by its decrease in the presence of b; and that the increase in B which takes place in the presence of A shall be exactly counterbalanced by its decrease in the presence of a. But it is important to notice that the equilibrium of the frequencies of the gametic combinations AB, Ab, aB, ab requires a third condition of equilibrium. By the conditions of our problem, two of these, which we have chosen to be AB and ab, are favoured by Natural Selection, and increase in their zygotic stages, while the opposite pair Ab and aB decrease. The adjustment of the ratio between the frequencies of these two pairs of gametic types must take place by recombination in those individuals which are heterozygotes for both factors. Of these so-called double heterozygotes some arise by the union of the gametic types AB and ab, and in these the effect of recombination is to diminish the frequencies of these two types. This effect will be partially counteracted by recombination in heterozygotes of the second kind, arising from the union of Ab and aB; and, if the net effect of recombination is to decrease the frequencies of AB and ab, it is obvious that double heterozygotes derived from gametes of these kinds must be the more numerous.

The inequality in the frequencies of the two kinds of double heterozygotes in the case we are considering has an important consequence; for whenever the two factors considered happen to be located in the same chromosome the frequency of recombination will depend upon crossing over, which is known to be much affected by the genetic differences between different strains. Moreover in the more numerous kind of double heterozygote recombination results in the substitution of the less favoured gametic combinations for the more favoured combinations, and consequently in a reduction in reproductive value, and this will not be completely balanced by the increase in reproductive value due to recombination in the less numerous kind of double heterozygote. Consequently the presence of pairs of factors in the same chromosome, the selective advantage of each of which reverses that of the other, will always tend to diminish recombination, and therefore to increase the intensity of linkage in the chromosomes of that species. This tendency is always in the same direction, and although the type of factorial interaction from which it arises may be rare, yet owing to the stability of the gene-ratios which it induces, we may anticipate that such cases will be found present at any one time with a frequency quite disproportionate to their rate of occurrence.

The discovery of an agency which tends constantly to increase the intensity of linkage, naturally stimulates inquiry as to the existence of other agencies having an opposite effect, and under the combined action of which, with that already discussed, linkage intensity could have become adjusted to its observed value. Such an agency appears to be at hand in the constant spread of advantageous mutations through the populations in which they occur. For, unless advantageous mutations occur so seldom that each has had time to become predominant before the next appears, they can only come to be simultaneously in the same gamete by means of recombination. If two advantageous mutations, which happen to be located in homologous chromosomes, are spreading simultaneously through the same species, we may look forward to a future epoch in which every gamete will contain both advantageous mutants, and these will have been derived in lineal succession, either from gametes of the same kind or, ultimately, from individuals in which recombination has taken place. Such individuals, we may infer, will have been, for this reason, somewhat better represented in future generations than the remainder, in which recombination frequency must have been, on the average, lower. It is apparent that for this process to have been an effective check upon the constant tendency to increase the intensity of linkage, the stream of favourable mutations must be an abundant one. There seems, however, to be no evidence against the view that even in every chromosome of most species numerous favourable mutations are at any one time always to be found, each as it increases in frequency, adding, perhaps only a trifle, to the perfection of its internal or external adaptation. If the need of combining these advantages is in reality the effective check to linkage intensity, it may prove possible, as data become more abundant, to gauge in this way, at least roughly, the relative rates of improvement in different species.

Simple metrical characters

Characters which can be specified by a single measurement, such as human stature, the length of an individual bone or tooth, etc., have a special importance owing to the fact that they can be studied relatively easily by direct biometrical methods. As has been pointed out in Chapter II the idea of adaptation cannot be applied with its full force to such simple characters, considered in isolation; but each may nevertheless be supposed to possess an optimum value in relation to the existing state of the organism and its environment, which we may regard as nearly coincident with the mean value exhibited by the species. That this must be so is evident from the extreme rapidity with which such measurements are modified when selection is directed to this end. For example, it appears from the observed average statures of the offspring of parents of different heights that no extreme selection would be needed to increase or decrease the stature of a human population by one inch in each generation, and even with the long generations of Man, such a rate of change would transcend the largest observed racial differences, within a short historical period.

If we consider any factor which affects such a measurement, and of which the other effects, if any, have no appreciable influence on survival, it is evident that the stability of its gene ratio requires separate consideration. For it would seem at first sight, if dominance were absent or incomplete, and in consequence the heterozygote were intermediate between the two homozygotes, that selection favouring intermediate values would tend to favour the heterozygotes, and in consequence induce very generally the condition of stability which has been considered. If this were so we should be faced with two somewhat alarming conclusions, (i) that by the accumulation in conditions of stability of a large number of factors with intermediate heterozygotes, the metrical characters should indicate in their bio-metrical properties a general absence of dominance, whereas, as has been already mentioned (p.18) the body of human measurements available give clear indications to the contrary, (ii) that the action of selection in favouring the intermediate values would have the effect, by preventing the extinction of all variant types, of progressively increasing the variance of the character in question, and consequently of making the intermediate values progressively rarer.

The recognition that the specific mean adjusts itself rapidly to the optimum size, however, makes the problem an essentially different one from that already considered, for the selective advantage of the heterozygote is dependent upon the average deviation of this genotype from the optimum, and this will vary as the gene ratio changes. If i, j, k represent the deviations of the average values of these genotypes from the mean, the effect of selection will be equivalent to eliminating small fractions of each genotype proportional to i^2, j^2, and k^2. If the three genotypes are in the proportion $p^2 : 2pq : q^2$, the ratios by which the two alternative genes are reduced will be proportional to

$$pi^2 + qj^2 \text{ and } pj^2 + qk^2,$$

and the gene ratio will only be in equilibrium if these two quantities are equal. Further from the definition of the mean, we have

$$p^2 i + 2pqj + q^2 k = 0.$$

If we use the latter equation to eliminate the deviations i, j, k, replacing them by a single ratio, defined as

$$\frac{a}{\beta} = \frac{i-j}{j-k},$$

and which depends only on the degree in which dominance is exhibited in the factor in question, we find that this ratio is connected with the ratio $p : q$, when the condition of equilibrium is established, by the equation

$$p(1 - 2p^2) a^2 + 2pq(q-p) a\beta - q(1 - 2q^2) \beta^2 = 0.$$

This expression is symmetrical if p and q, a and β are interchanged, the value on the left being proportional to the rate of decrease of $z\ (= \log p - \log q)$.

From the signs of the three terms it appears that one real positive solution exists, if q exceeds p, only if q does not exceed $\frac{1}{2}\sqrt{2}$; moreover since the expression is positive if $\beta = 0$, while if $a = \beta$ it is reduced to

$$(q-p)\beta^2$$

which is still positive, it follows that if q exceeds p, so must β exceed a, when equilibrium is attained. The more dominant gene must be the less frequent.

If any equilibrium were stable then the expression must increase as p is increased; its differential coefficient with respect to p is

$$(1 - 6p^2) a^2 + (1 - 6pq) 2a\beta + (1 - 6q^2) \beta^2,$$

or

$$(a + \beta)^2 - 6(pa + q\beta)^2.$$

Now $a + \beta$ is arbitrary, and may be taken like $p + q$ to be unity, then since

$$2(pa + q\beta) = (p + q)(a + \beta) + (p - q)(a - \beta)$$

it follows that $pa + q\beta$ cannot be less than $\frac{1}{2}$, for we have shown that when q exceeds p, then β exceeds a. Consequently the differential coefficient at any position of equilibrium is less than

$$1 - \frac{3}{2},$$

and is always negative. The conditions of equilibrium are always unstable. Whichever gene is at less than its equilibrium frequency will tend to be further diminished by selection.

All mutations therefore affecting such a character, unless they possess countervailing advantages in other respects, will be initially disadvantageous, and we may conceive of each coming to an equilibrium at which the mutation rate is just balanced by the counter-selection to which it is exposed. This situation resembles that of intrinsically disadvantageous mutations considered in Chapter III, but differs from it in that for sufficiently high mutation rates the mutant gene will now pass the point of maximum resistance, and, when it attains sufficient frequency, will be thereafter actually assisted by selection. The mutation rates required to bring this about depend on (i) the magnitude of the effect produced by the factor, i. e. the metrical difference between the two homozygotes, represented by $2a$, (ii) the total variance of the species in the measurement in question, represented by σ^2, (iii) the intensity of selection in favour of the optimum measurement; this may be measured by $1/\tau^2$, where τ^2 is a quantity of the same dimensions as σ^2, which vanishes for infinitely intense selection, and would be infinitely great if all values of the measurement were equally satisfactory.

In the absence of dominance the maximum resistance is encountered when $p = \frac{1}{4}$, and this is overcome if the mutation rate, k, exceeds

$$\frac{a^2}{16\,(\sigma^2 + \tau^2)}\,;$$

a full discussion of the situation would require an examination of the effect of selection upon the variance, for if selection is intense and τ^2 small, it is probable that σ^2 will become small also; for the present we may note that the contribution of the factor in question to the total variance will be at most $a^2/2$. If therefore the effect of the factor is so small that it will contribute at most one part in 100,000 to the total variance, a mutation rate of the order of one in a million might well effect its gradual establishment. Such would be the situation of factors affecting human stature by about one-fortieth of an inch. Factors having less effect than this might establish a mutant form at lower mutation rates, in each case the more easily, the more lax is the preferential survival of the medium sizes.

Mutant genes with greater effect or lower mutation rates will be hung up at all values up to $p = 0.25$. Now at this value the mean lies midway between the average values of the heterozygote and the non-mutant homozygote; consequently at and below this value the heterozygote might with advantage more nearly resemble the non-mutant homozygote. There will, therefore, always be a tendency, analogous to that discussed in Chapter III, for the non-mutant gene to become dominant, by the modification of the heterozygote towards greater resemblance with it. Dominance should then be developed against mutations in whichever direction they appear; thus we may expect to find in such characters mutant genes of which the tendency is either to increase or to decrease the measurement, indiscriminately recessive. The fact that the offspring of crosses between races exhibiting differences in metrical characters are usually intermediate is one which might have been inferred from this bilateral tendency towards the development of dominance.

The development of dominance inevitably reacts upon the conditions of equilibrium; for complete dominance, for example, the maximum counterselection is met with at $p = 0.5$, and the mutation rate necessary to establish the mutant gene is four times as great as the value found for factors without dominance. The condition necessary for dominance to increase is that the heterozygote shall be on the opposite side of the mean to the non-mutant homozygote. For any particular degree of dominance, that is for any particular value of a, dominance will increase so long as

$$\frac{a}{\beta} < \frac{q^2}{p^2}.$$

With sufficient time, therefore, dominance will increase, and the value of β diminish, until these two values are equal. The simultaneous variation of the two ratios $p : q$ and $a : \beta$ will be made more clear by the aid of the diagram (Fig. 9) on which the curve $AXZB$ represents the series of possible conditions in which there is no further tendency to modify the degree of dominance, and at different points along which factors may be maintained by appropriate mutation rates. On the same diagram the line $EAYC$ is drawn through the points at which maximum counterselection is met with for different values of a. A mutation commencing without dominance will start from O, and as its gene frequency increases, move along the line OE, until it reaches some point on this line at which the mutation rate is balanced by counterselection. At this stage, and indeed during its progress towards this stage, selection will tend to render the mutant gene recessive and the representative point will pass along a line below EZ to come to rest on the limiting line ZB, at all points of which, as appears from the diagram α exceeds 0·95.

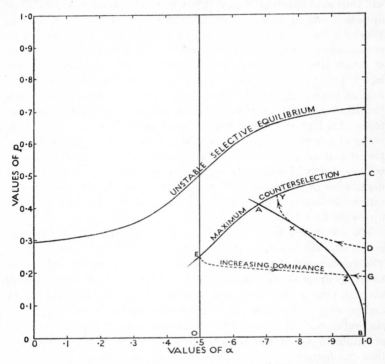

FIG. 9. The relations between frequency, mutation rate, and degree of dominance, in a factor having simple metrical effect only. For detailed explanation see text.

At any stage an alteration in the mutation rate will move the representative point upwards or downwards on the diagram according as the mutation rate is increased or diminished, but it is clear that a factor left without change in the area $AEOB$ will gravitate to the appropriate level on the line XB. In the same way any factor displaced to a point in the area XDB will experience selection in the direction of a diminution of dominance, for with these the displacement of the heterozygote from the mean of the species is in the same direction as that of the non-mutant homozygote and selection will favour any tendency to diminish this deviation. Such a path is represented by GZ. From the position of the point X it will be seen that this process cannot reduce the value of a below about 0·79 unless the critical line AC is crossed. Factors held in equilibrium will thus tend to retain at least this degree of dominance, even if their mutation rate is very nearly sufficient to overcome all opposition. Finally a factor placed in the region $CYXD$ will, if the conditions remain unchanged, diminish its dominance until it meets the line AC, at which the effect of counterselection is a maximum and beyond which it grows weaker instead of stronger, when the frequency of the mutant gene is increased. The course of such factors is, with some vicissitudes, similar to that of factors which, combining the advantages of a high mutation rate and a very small measurable effect, never encounter sufficient opposition to check their progress. The lowest mutation rate which could maintain the factor on such a path is very nearly double that initially required for a factor without dominance to pass at E, and is represented by the line DXY separating the factors which will pass AC from those which will be hung up on the line XZB.

We are in the dark as to the frequencies with which different mutations will attain the different stages indicated in the diagram. The greater part of the area represented could, however, only come to be occupied by mutations which fail by but little from passing through unopposed. Mutations having a comparatively large effect and possibly many others with less effect, combined with low mutation rates, may be expected to be checked at low values of p and to gravitate to points near B on the line ZB, where it will be observed that dominance, while still slightly incomplete, attains a very pronounced development.

A second peculiarity of metrical factors, and one which may be of more consequence, is that, of any two genes having similar effect, that is both increasing or both diminishing the measurement, each will be most advantageous or least disadvantageous, in the absence of the other. The preceding analysis has shown that this interaction does not lead to stability of the gene ratios, comparable to that discussed on p. 102; nevertheless the interaction must have exactly analogous effects in favouring genotypes which exhibit close linkage. Gametic combinations in which the different factorial effects are well balanced and tend to a measurement of medium length, must be continually favoured by selection at the expense of ill-balanced combinations determining the production of very large or very small values. The numbers of the latter will therefore require on the average to be continually replenished by recombination, with the result that crossing-over must, so far as these factors are concerned, tend to lower the average reproductive value of the offspring.

It will now be clear in what way we should imagine the average value of the measurements to be modified by selection whenever such modification happens to be advantageous. If the optimum value is increased all genes, the effect of which in contrast to their existing allelomorphs is a metrical increase, will be immediately, or at least rapidly, increased in frequency. In the case of those factors, the effects of which upon survival can be completely expressed in terms of their effect upon the measurement in question, the effects of such a change of frequency will be in some cases permanent and in others temporary. Some genes previously opposed by selection will be shifted to frequencies at which they are favoured, and these may increase to such an extent during the period of selection that when this dies away they may still be favoured. Their subsequent progress will thus tend to increase the value of the measurement even after it has attained the new optimum. The same applies to mutant genes the frequency of which is increased past the point of maximum counterselection, into a region in which selection, while still opposing mutation, is insufficient to check their increase. In other cases the increase or decrease in frequency produced by temporary selection, is itself only temporary, and these, when the new optimum is attained, will tend to revert to their previous frequencies, tending incidentally by so doing, to make the specific mean revert somewhat from the new optimum. The system resembles one in which a tensile force is capable of producing both elastic and permanent strain, and in which the permanent deformations always tend to relieve the elastic forces which are set up.

Meristic characters

In his search for evidences of discontinuous variation Bateson paid considerable attention to variations in the number of similar parts occurring, like vertebrae, in series. Apart from abnormalities in development, variation in such characters is bound to be discontinuous, the variate exhibited by any one individual being necessarily one of the series of whole numbers. There is no reason, however, to suppose that the discontinuity so produced is in any way connected with the discontinuity of the genetic particles in Mendelian inheritance. On the contrary, it would be equally reasonable to suppose *a priori*, that the actual number exhibited is but the somatic expression, to the nearest whole number, of an underlying physiological variate influenced, like a simple measurement, by both environmental and genetic causes. That this is the true view must now be regarded as established in several important cases.

The series of frequencies with which different vertebra numbers occur bear a striking superficial resemblance to the series obtained when a normally distributed variate is grouped in equal but somewhat large intervals of its value; as if, for example, human stature were recorded in a large number of individuals, to the nearest multiple of three inches. The impression of similarity is increased if instead of observations upon single individuals, we consider the simultaneous distribution of a number of pairs of parents and offspring, for in this case the average value of the meristic variate in a group of offspring obtained from parents with the same value, is found to increase quantitatively from group to group in a manner exactly similar to that observed with simple measurements.

The actual proof that each individual possesses, in respect of a character showing meristic variation, a definite genotypic or genetic value, which differs from the corresponding values of other individuals by amounts which are not integers, but fractions differing significantly from integers, can, as Schmidt has shown, be supplied by two methods of experimentation. With organisms capable of self-fertilization, or vegetative reproduction, it may be possible to establish pure lines of individuals genotypically identical, and a direct comparison can thus be made between the averages of large numbers of individuals of two or more different lines, developed in the same environment. Alternatively with organisms capable of giving a sufficient number of offspring at a single mating, a series of males may be bred, each to every one of a series of females. By comparing the averages of these progenies Schmidt determined the differences between the underlying genetic values of parents of the same sex, with sufficient precision to show that these differences could not possibly be integral.

If we regard such a variate as vertebra number as the somatic expression of an underlying genotypic variate having continuous variation, the striking constancy of such meristic characters in large groups of related organisms evidently requires a special explanation. This constancy is made no less remarkable by its exceptions. Among the great diversity of forms developed among the mammalia, almost all have constantly seven neck vertebrae, yet two of the sloths have six and nine neck vertebrae respectively. A similar situation in fishes, in which several families with various vertebra numbers have apparently developed at different times, from a group of families in which 24 vertebrae appear to be invariable, has been felt by Tate Regan to present such a difficulty to the theory of Natural Selection, that he is willing to fall back upon the supposed effects of changed conditions in producing mutations as an alternative explanation. What prospect there is of such an agency, if its existence could be demonstrated, aiding us in understanding this particular problem seems at present uncertain. It is therefore the more important to examine whether known causes are really as ineffective as has been thought in bringing about the observed effects.

In groups in which all or nearly all the individuals have the same vertebra number two views are possible; (i) that there is no genetic variability, and (ii) that neither genetic variability, nor the variability of the developmental environment, is sufficient to produce frequent departures from the central integer. The first view may be set aside, not only because different species do certainly differ in the number of their vertebrae, but also because, in the light of the argument of the last chapter, a mutant gene affecting the underlying variate, unless it have other effects, will be exempt from selection, at least so long as the vertebra number is actually constant. Consequently, any mutations of this kind which have occurred in the past must accumulate in such species.

If we take the second view, heritable individual variation exists in respect of the tendency to produce a given number of vertebrae, and the species is therefore potentially plastic in this respect. Supposing the mean of this distribution to coincide with the modal integer, the frequency of values other than this integer may be easily calculated from the standard deviation of the distribution; for example, if the standard deviation is $\frac{1}{6}$ of a unit, about three exceptions are to be expected among a thousand individuals, for $\frac{1}{8}$ of a unit only sixty-three in a million, for $\frac{1}{10}$ of a unit only one in two million, and so on.

Very extensive counts would therefore be required to exclude variation of these amounts, which would nevertheless be sufficient to permit
of an evolutionary change in vertebra number, if at any time this
became advantageous.

The evolutionary fact, which on this view requires a special explanation, is that in large groups of organisms with widely diverse
adaptations, it has so seldom been found advantageous to make such
a change, for in this the meristic variates offer a very striking contrast to the metrical variates. The integral numbers appear to be
possessed of a special kind of stability favouring a conservative
tendency in evolution, which is not to be found in the simple measurements. It is possible that the explanation of this tendency lies in
the simple fact that the intercalation or omission of a member of
a series of structures requires a corresponding modification of a
number of associated organs, such as attached muscles, nerves, blood-
vessels, etc.; and that even in cases where there is a slight advantage
to be gained by a complete reorganization on the basis of an increased
number of vertebrae, it may well be that such advantage is less than
the disadvantage suffered initially owing to the disorganization of
associated structures in any individuals which happen to have the
higher number. Even if the associated structures were morpho-
logically complete, it is not certain that all quantitative physiological
adjustments will be perfectly co-ordinated. The argument is a special
case of a more general one to the effect that in any highly adapted
organism the probability of advantage through any considerable
evolutionary step (saltation) rapidly becomes infinitesimal as the
step is increased in magnitude.

The liability to maladjustment should of course be least in species
showing considerable variability, such as the eel, in which we may
expect the developmental processes to be carried through nearly
perfectly whatever the actual number of vertebrae laid down. It
should be greatest where the meristic variate is most constant, and
where the capacity of the developmental mechanisms for dealing
with other numbers has not been subjected to selection in previous
generations. In these, however, the rarity of the exceptions precludes
the possibility of demonstrating any associated signs of abnormal
development. An intermediate case, for which some data are available, is afforded by the herring. Ford and Bull have studied the
occurrence of fused and abnormal vertebrae in this fish. Out of

nearly 7,000 skeletons examined 95 were found to contain abnormal or multiple structures. If each element in these structures is counted as a whole vertebra it is found that the mean vertebra number of the abnormal skeletons is 55·82, which exceeds the mean value for normal skeletons by only 0·03. Although the mean values are thus brought into close agreement the variability of the abnormal skeletons is very much greater than that of the normal. In other words the extreme vertebra numbers 53 and 58 show the highest percentages of abnormal skeletons, the median vertebra numbers 55 and 56, which contain nearly 90 per cent. of the fish, show the lowest percentages, while the intermediate numbers 54 and 57 show intermediate percentages abnormal. The actual percentages obtained from Ford and Bull's data are as follows:

TABLE 6.

Vertebra number.

	53	54	55	56	57	58
Per cent. abnormal . .	45·5	4·1	1·2	1·1	2·6	10·0
Percentage frequency .	0·08	1·06	28·36	61·30	8·91	0·29

It will be observed that the fish with the rarer vertebra numbers show a very pronounced liability to develop abnormal structures. If such a tendency is general in the case of unwonted meristic variations, as seems on general grounds to be extremely probable, any meristic variation from the existing standard must in general encounter appreciable counter-selection, and could scarcely establish itself unless the increased number conferred advantages which could be obtained in no other way.

If we are right in referring the conservative tendency, which is observed to hold in meristic matters, to the liability to disorganization consequent upon a sudden change, it follows that this tendency must be in at least partial abeyance at periods in which the associated structures, or their relationships to the vertebrae are, for other reasons, in a state of reorganization. Such a pronounced change in the structure and movements of the body as must have been required in the adoption of the habits of the flat fish, or in the musculature of the neck by the inverted position of the sloth, may thus have afforded a temporary opportunity for a form of variability always present, though constantly kept in subjection, to produce for once an evolutionary modification.

Biometrical effects of recent selection

The immediate effect of selection in either direction is to change the average value of the metrical or meristic variate, as the case may be. Such an effect can only be detected if successive average values can be determined, as is the case with some palaeontological material. A further effect may be anticipated in a slight reduction of the variance, by the elimination of factors previously contributing to it; this effect will be extremely small, and would in any case give no indication as to whether the selection had favoured an increase or decrease in size. Among the third degree functions of the measurements there should, however, be traces of recent selection, which would differ according to the direction in which the selection has been applied. For a selection in favour of increased size would only temporarily displace recessive genes having a dwarfing effect, and these should remain in the species during such a process. On the other hand a recessive gene, the effect of which is to increase the metrical character, would have a chance during such a process of becoming established throughout the species, by the extinction of its allelomorph. Consequently after such selection has been long in action the recessive genes should more frequently tend to diminish than to increase the character studied. Equally when, as in the development of toy breeds of dogs, selection has favoured diminished size, we should expect to find an excess of recessives tending to increase the average dimensions.

The biometrical detection and measurement of any prevalence of dominance in one direction over that in the other would certainly require very ample material, but apart from this the necessary calculations appear to be straightforward. The simplest method, that of measuring the asymmetry of the frequency distribution by means of the third moment, will encounter some difficulties of interpretation owing to the effects of environmental variations; for this reason it may be inferior to the study of the asymmetry of large fraternities derived from different pairs of parents. The most direct investigation would be rendered possible by considering in conjunction the measurements of father, mother, and offspring, and expressing the regression of the size of the offspring upon the sizes of the two parents in a regression formula of the form

$$ax + by - cxy$$

in which, x and y are the parental measurements, and a, b, c the coefficients to be determined from the data. Selection for increased size will then tend to produce positive values of c, or in general to increase its value algebraically, while selection in the opposite direction will tend to decrease it, and to produce negative values.

Summary

In the two preceding chapters an attempt has been made to check the adequacy of the genetic theory of Natural Selection by a detailed examination of the method by which on this view new genes arise, prevail, or become extinct, and in particular new gene contrasts, or Mendelian factors, become temporarily or permanently established as contributors to that stock of organic variability from which the rate of progress of species has been shown to depend.

The examination of the important problem of the survival of individual genes is much simplified by the fact that so long as few are in existence all other effects are unimportant compared with the fortuitous element in the survival and reproduction of their bearers. In many important cases the survival for a single generation may be represented by a function simply related to the Poisson series, and in many other cases by the substitution for a generation of the appropriate biological cycle the same formulae will supply an excellent approximation.

Although initially subject to the full force of random survival, beneficial mutations have a finite probability, simply related to the benefit which they confer, of establishing themselves as permanent in the heredity of the species. They can, therefore, have occurred but a small number of times before this event is rendered practically certain. The mutation rates during their period of trial must therefore be generally exceedingly minute, and it must frequently happen that nearly the whole of the individual genes that ultimately pervade the species will have been derived by descent from a single such mutant.

An examination of the statistical equilibrium established between new mutations and the causes of the extinction of genes, shows that advantageous mutations, at all levels of advantage, resemble neutral mutations in the distribution of gene ratios established, the frequency being equal in all equal intervals of the variate z, in which it is convenient to measure variations in gene ratio. The range of z is longer in abundant than in rare species, and for this reason the number of factorial differences maintained in an abundant species bears a higher ratio to the rate of occurrence of new mutations than does the number maintained in rare species. The range of selective advantage which may be regarded as effectively neutral is, however, extremely minute, being inversely proportional to the population of the species Since it is scarcely credible that such a perfect equipoise of selective advantage could be maintained during the course of evolutionary change, random survival, while the dominant consideration in respect to the survival of individual genes, is of merely academic interest in respect to the variance maintained in the species, which must be mainly supplied by definitely advantageous mutations. In particular the rate of decay of variance due to random extinction in the total absence of mutations is shown to be of the most trifling importance.

There are, on the theory here developed, the strongest theoretical grounds for concluding that the more numerous species must *ceteris paribus* tend to be the more variable, though the rate at which variability should increase in relation to increased numbers can, it would seem, only be established by direct observation. The relationship between abundance and variability has been confirmed by a fine body of observations on moths, but deserves extensive investigation in other groups. An evolutionary consequence of some importance is that in general a smaller number of large species must be increasing in numbers at the expense of a larger number of small species, the continuous extinction of the latter setting a natural check to the excessive subdivision of species which would ensue upon a too fine and detailed specialization.

Of great importance for our subject is the occurrence, even when due to exceptional circumstances, of gene ratios which are stable under selective influences; since factors of this kind cannot be eliminated unless and until, in the process of evolutionary change, the stability is upset. The simplest type of these occurs when the heterozygote of a single factor is favoured by selection at the expense of both homozygotes. It is shown that in this case an equilibrium always exists and is always stable, whereas in the opposite case the equilibrium is unstable, and the less numerous gene will be continuously eliminated until its extinction. The accumulation of factors in which the heterozygote is favoured will give a constant average advantage to cross- over self-fertilization, in the sense that the progeny by cross-fertilization has a higher average reproductive value, and is worth more to produce, than the progeny by self-fertilization. The development of separate sexes in motile animals and of the many devices to ensure cross-pollination in plants, though not the origin of sexual reproduction itself, seems to be ascribable to the small constant individual advantages due to the favouring of the heterozygote ; at least if we include in this phrase the constant tendency of the heterozygote to resemble the more favourable homozygous form, examined in Chapter III.

Other types of stable equilibrium may be established by the interaction of two or more factors. The importance of this group of cases lies in the constant tendency, whenever two such factors are in the same linkage group, for the linkage between them to be increased. The same consequence ensues, without the establishment of stable equilibrium, when two factors of such a group both modify the magnitude of any simple physical measurement, and have no other effects of importance to survival. There is thus shown to be an agency constantly favouring closer linkage between factors of the same group, and since linkage values are eminently liable to selective modification, and are not, in most species investigated, such as to preclude frequent crossing over, it is an inevitable inference that some other cause must induce an equally powerful selection in favour of crossing over. While the mathematical difficulties of an exact investigation are worthy of a far more extended treatment, it is suggested that such an agency may be found in the advantage of combining different advantageous mutations, which, unless they occur consecutively, can only be done by recombination. It seems probable that in order to exert a perceptible influence upon linkage the stream of favourable mutations would need to be a considerable one, and that with further extensions of our knowledge in this direction it may prove possible, by this means, to gauge its magnitude.

Mutations whose effect is produced only upon simple metrical characters will, unless their effects are very minute, be exposed to counter-selection, and the mutant gene, whether its effect be positive or negative, will tend to become recessive. Dominance in such cases should be incomplete, though in most factors exposed long to counter-selection the heterozygote should differ from the non-mutant homozygote by less than 1/20 of the difference between the homozygotes. The evidence of dominance observed in metrical characters is thus in full accord with the analytic theory.

Meristic variation is, in those cases which have been investigated, certainly due to an underlying quantitative variate. Specific modification in such characters is, however, constantly opposed by selection, arising by the interaction of co-ordinated structures. The striking conservatism manifested by large groups of organisms in meristic characters is thus rendered intelligible, while the possibility of their modification at any time, during the reorganization of the relationships and attachments of associated structures, is constantly maintained. Such anomalies as occur in the numerical variations in the neck vertebrae of mammals may be cited as illustrating both sides of the working of this principle.

VI

SEXUAL REPRODUCTION AND SEXUAL SELECTION

The contrast between sexual and asexual reproduction. The nature of species. Fission of species. Sexual preference. Sexual selection. Sex limitation of modifications. Natural Selection and the sex ratio. Summary.

To all who are engaged in Psyche's task, of sorting out the seeds of good from the seeds of evil, I dedicate this discourse. FRASER.

The contrast between sexual and asexual reproduction

A GROUP of organisms in which sexual reproduction was entirely unknown might none the less evolve under the action of natural selection. This condition cannot, I believe, be ascribed with certainty to any known group. Yet, since it is impossible to draw any sharp distinction within a whole series of asexual processes, from individual growth at the one extreme, through the regeneration of injured or lost parts, to vegetative reproduction by budding; it is tempting to believe that asexual reproduction was the primitive condition of living matter, and that the sexual reproduction of the predominant types of organisms is a development of some special value to the organisms which employ it. In such an asexual group, systematic classification would not be impossible, for groups of related forms would exist which had arisen by divergence from a common ancestor. Species, properly speaking, we could scarcely expect to find, for each individual genotype would have an equal right to be regarded as specifically distinct, and no natural groups would exist bound together like species by a constant interchange of their germ-plasm.

The groups most nearly corresponding to species would be those adapted to fill so similar a place in nature that any one individual could replace another, or more explicitly that an evolutionary improvement in any one individual threatens the existence of the descendants of all the others. Within such a group the increase in numbers of the more favoured types would be balanced by the continual extinction of lines less fitted to survive, so that, just as, looking backward, we could trace the ancestry of the whole group back to a single individual progenitor, so, looking forward at any stage, we can foresee the time when the whole group then living will be the descendants of one particular individual of the existing population. If we consider the prospect of a beneficial mutation occurring at any instant, ultimately prevailing throughout the whole group, and so leading to evolutionary progress, it is clear that its prospect of doing so will depend upon its chance of falling, out of the whole population, upon the one individual whose descendants are destined ultimately to survive. At first sight this chance appears to be extremely small; but we must take account of the fact that in so far as the mutation is beneficial, its occurrence will increase the prospect of the individual, in which it occurs, proving ultimately victorious. In the limiting case in which the benefit derived from the new mutation tends to zero the chance of success is evidently only one in as many individuals as there are in the competing group. If on the other hand the benefit is appreciable, the chance of success will certainly be greater than this by an amount which now depends on the amount of heritable diversity in the group, and on the prospect of the occurrence of other beneficial mutations, before the replacement of the original population by the improved type has been completed. If the total rate of mutations is so small that the usual condition of the group is one of genetic uniformity, any advantageous mutation may be expected to prevail, provided it survives the chances of accidental death during the initial period in which it is represented by only one or few individuals. These chances, which are effectively the same with asexual or with sexual reproduction, have been considered in an earlier chapter (IV).

The evolutionary progress of an asexual group thus presents the dilemma that it can only utilize all those beneficial mutations which occur, and survive the dangers of the initial period, if the rate of occurrence of mutations is so low that the population of competing organisms is normally in a state of genetic uniformity, and in such a state evolutionary progress will necessarily be almost at a standstill; whereas if on the contrary the mutation rates, both of beneficial and of deleterious mutations, are high enough to maintain any considerable genetic diversity, it will only be the best adapted genotypes which can become the ancestors of future generations, and the beneficial mutations which occur will have only the minutest chance of not appearing in types of organisms so inferior to some of their competitors, that their offspring will certainly be supplanted by those of the latter. Between these two extremes there will doubtless be an optimum degree of mutability, dependent on the proportion of beneficial to deleterious mutations, and therefore on the aptitude of the group to its place in nature; but it is not difficult to see that the rate of progress, supposing that the optimum mutability were established, would still be very inferior to that of a sexual organism placed in the same circumstances.

The argument developed above as to the rate of evolutionary progress of a group of asexual organisms may be applied to the evolutionary progress in any one particular locus, in a species of sexual organisms, if we suppose that changes of several different kinds may take place in an homologous set of genes. The comparative rates of progress of sexual and asexual groups occupying the same place in nature, and at the moment equally adapted to that place, are therefore dependent upon the number of different loci in the sexual species, the genes in which are freely interchangeable in the course of descent. From what is known of the higher animals this number must be at least several thousands; but even a sexual organism with only two genes would apparently possess a manifest advantage over its asexual competitor, not necessarily from any physiological benefit derived from sexual union, but from an approximate doubling of the rate with which it could respond to Natural Selection. On this view, although asexual reproduction might be largely or even exclusively adopted by particular species of sexual groups, the only groups in which we should expect sexual reproduction never to have been developed, would be those, if such exist, of so simple a character that their genetic constitution consisted of a single gene.

The nature of species

From genetic studies in the higher organisms it may be inferred, that whereas genetic diversity may exist, perhaps in hundreds of different loci, yet in the great majority of loci the normal condition is one of genetic uniformity. Unless this were so the concept of the wild type gene would be an indefinite one. Cases are indeed known, as in the agouti locus in mice, in which more than one kind of wild gene have been found, these being both dominant to their other non-lethal allelomorphs; but numerous as are the loci in which such genetic diversity must exist, we have some reason to suppose that they form a very small minority of all the loci, and that the great majority exhibit, within the species, substantially that complete uniformity, which has been shown to be necessary, if full advantage is to be taken of the chances of favourable mutations. In many loci the whole of the existing genes in the species must be the lineal descendants of a single favourable mutation.

The intimate manner in which the whole body of individuals of a single species are bound together by sexual reproduction has been lost sight of by some writers. Apart from the intervention of geographical barriers so recently that the races separated are not yet regarded as specifically distinct, the ancestry of each single individual, if carried back only for a hundred generations, must embrace practically all of the earlier period who have contributed appreciably to the ancestry of the present population. If we carry the survey back for 200, 1,000, or 10,000 generations, which are relatively short periods in the history of most species, it is evident that the community of ancestry must be even more complete. The genetical identity in the majority of loci, which underlies the genetic variability presented by most species, seems to supply the systematist with the true basis of his concepts of specific identity or diversity. In his *Materials for the Study of Variation*, W. Bateson frequently hints at an argument, which evidently influenced him profoundly, to the effect that the discontinuity to be observed between different species must have owed its origin to discontinuities occurring in the evolution of each. His argument, so far as it can be traced from a work, which owed its influence to the acuteness less of its reasoning than of its sarcasm, would seem to be correct for purely asexual organisms, for in these it is possible to regard each individual, and not merely each specific type, as the last member of a series, the continuity or discontinuity of which might be judged by the differences which occur between parent and offspring; and so to argue that these provide an explanation of the diversity of distinct strains. In sexual organisms this argument breaks down, for each individual is not the final member of a single series, but of converging lines of descent which ramify comparatively rapidly throughout the entire specific group. The variations which exist within a species are like the differences in colour between different threads which have crossed and recrossed each other a thousand times in the weaving a single uniform fabric.

The effective identity of the remote ancestry of all existing members of a single sexual species may be seen in another way, which in particular cases should be capable of some quantitative refinement. Of the heritable variance in any character in each generation a portion is due to the hereditary differences in their parents, while the remainder, including nearly all differences between whole brothers and sisters, is due to genetic segregation. These portions are not very unequal; the correlations observed in human statistics show that segregation must account for a little more than two-fifths, and the hereditary differences of the parents for nearly three-fifths of the whole. These hereditary differences are in their turn, if we go back a second generation, due partly to segregation and partly to hereditary differences in the grandparents. As we look farther and farther back, the proportion of the existing variance ascribable to differences of ancestry becomes rapidly smaller and smaller; taking the fraction due to segregation as only $\frac{2}{5}$ in each generation, the fraction due to differences of ancestry 10 generations back is only about one part in 160 while at 30 generations it is less than one in four millions. It is only the geographical and other barriers to sexual intercourse between different races, factors admittedly similar to those which condition the development of incipient species as geographical races, which prevent the whole of mankind from having had, apart from the last thousand years, a practically identical ancestry. The ancestry of members of the same nation can differ little beyond the last 500 years; at 2,000 years the only differences that would seem to remain would be those between distinct ethnographic races; these, or at least some of the elements of these, may indeed be extremely ancient; but this could only be the case if for long ages the diffusion of blood between the separated groups was almost non-existent.

Fission of species

The close genetic ties which bind species together into single bodies bring into relief the problem of their fission—a problem which involves complexities akin to those that arise in the discussion of the fission of the heavenly bodies, for the attempt to trace the course of events through intermediate states of instability, seems to require in both cases a more detailed knowledge than does the study of stable states. In many cases without doubt the establishment of complete or almost complete geographical isolation has at once settled the line of fission; the two separated moieties thereafter evolving as separate species, in almost complete independence, in somewhat different habitats, until such time as the morphological differences between them entitle them to 'specific rank'. It would, however, be contrary to the weightiest opinions to postulate that specific differentiation had always been brought about by geographic isolation almost complete in degree. In many cases it may safely be asserted that no geographic isolation at all can be postulated, although this view should not be taken as asserting that the habitat of any species is so uniformly favourable, both to the maintenance of population, and to migration, that no 'lines of weakness' exist, which, if fission is in any case imminent, will determine the most probable geographic lines of division. It is, of course, characteristic of unstable states that minimal causes can at such times produce disproportionate effects; in discussing the possibility of the fission of species without geographic isolation, it will therefore be sufficient if we can give a clear idea of the nature of the causes which condition genetic instability.

Any environmental heterogeneity which requires special adaptations, which are either irreconcileable or difficult to reconcile, will exert upon the cohesive power of the species a certain stress. This stress will be least when closely related individuals are exposed to the environmental differences, and vanishes absolutely if every individual has an equal chance of encountering either of two contrasted environmental situations, or each of a graded series of such situations. It is greatest when associated with circumstances unfavourable to sexual union, of which the most conspicuous is geographical distance, though others, such as earliness or lateness in seasonal reproduction, may in many cases be important. I do not know any such circumstance, which, in the genetical situation produced, differs essentially from geographical distance, in terms of which, therefore, it is convenient to develop the theory.

We may consider the case of a species subjected to different conditions of survival and reproduction at opposite ends of its geographical range. Certain of the genes which exist as alternatives will be favoured at one extreme, and will tend there to increase, while at the other extreme they will be disadvantageous and tend to diminish in frequency, the intermediate region being divisible into a series of zones in which the advantage increases, from a negative value at one extreme, through zero at a region in which the selective advantage is exactly balanced, to a certain positive advantage at the other extreme. A condition of genetic equilibrium is therefore only established if the increase in frequency in the favourable region and the decrease in frequency in the unfavourable region, not only balance each other quantitatively, but are each equal to the rate at which genes diffuse by migration and sexual union, from the one region to the other. This rate must itself be determined, apart from migratory or sedentary habits of the species, by the length of each zone across which diffusion occurs, by the density of population along it, and finally by the gradient in the frequency ratio between the gene and its allelomorph as we pass across it. So long as a sufficient gradient can be maintained, accompanied by an active diffusion of germinal material, so long the local varieties, although, possibly, distinct differences between them may be detected, will have no tendency to increase these differences in respect of the frequency of the genes in which they differ, and will be connected by all grades of intermediate types of population.

The longer such an equilibrium is maintained the more numerous will the genetic differences between the types inhabiting extreme regions tend to become, for the situation allows of the extinction of neither the gene favoured locally nor its allelomorph favoured else-where, and all new mutations appearing in the intermediate zone which are advantageous at one extreme but disadvantageous at the other will have a chance of being added to the factors in which they differ. In addition to those genes which are selected differentially by the contrasted environments, we must moreover add those, the selective advantage or disadvantage of which is conditioned by the genotype in which they occur, and which will therefore possess differential survival value, owing not directly to the contrast in environments, but indirectly to the genotypic contrast which these environments induce. The process so far sketched contains no novel features, it allows of the differentiation of local races under natural selection, and shows that this differentiation must, if the conditions of diffusion are constant, be progressive. It involves no tendency to break the stream of diffusion, or consequently to diminish in degree the unity of ancestry which the species possesses. It is analogous to the stretch-ing of a material body under stress, not to its rupture.

There are, however, some groups of heritable variations which will influence diffusion. In the case we are considering in which the cause of isolation is geographical distance, the instincts governing the movements of migration, or the means adopted for dispersal or fixa-tion, will influence the frequency with which the descendants of an organism, originating in one region, find themselves surrounded by the environment prevailing in another. The constant elimination in each extreme region of the genes which diffuse to it from the other, must involve incidentally the elimination of those types of individuals which are most apt so to diffuse. If it is admitted that an aquatic organism adapted to a low level of salinity will acquire, under Natural Selection, instincts of migration, or means of dispersal, which minimize its chances of being carried out to sea, it will be seen that selection of the same nature must act gradually and progressively to minimize the diffusion of germ plasm between regions requiring different specialized aptitudes. The effect of such a progressive diminution in the tendency to diffusion will be progressively to steepen the gradient of gene frequency at the places where it is highest, until a line of distinction is produced, across which there is a relatively sharp contrast in the genetic composition of the species. Diffusion across this line is now more than ever disadvantageous, and its progressive diminution, while leaving possibly for long a zone of individuals of intermediate type, will allow the two main bodies of the species to evolve almost in complete independence.

In cases in which the cause of genetic isolation is not merely geographical distance, but a diversity among different members of the species in their habitats or life history, in connexion with which different genetic modifications are advantageous; the isolation will of course not be increased by the differential modification of the instincts of migration, or the means of dispersal; but by whatever type of hereditary modification will minimize the tendency for germinal elements, appropriate to one form of life, to be diffused among individuals living the other form, and among them consequently eliminated.

The power of the means of dispersal alone, without the necessity for selective discrimination in either region, is excellently illustrated by the theory, due to Ray Lankester, which satisfactorily accounts for the diminution or loss of functional eyes by the inhabitants of dark caverns. Ray Lankester pointed out that the possession of the visual apparatus is not merely useless to such animals but, by favouring their migration towards sources of light, will constantly eliminate them from the body of cave inhabitants, equally effectively whether they survive or perish in their new environment. Those which remain therefore to breed in the cavern are liable to selection in each generation for their insensibility to visual stimuli. It should be noted that with such very restricted habitats migrational selection of this sort might attain to very high intensity and in consequence produce correspondingly rapid evolutionary effects.

Sexual preference

A means of genetic isolation which is of special importance in that it is applicable equally to geographical and to other cases is one, which for want of a better term, we may consider under the heading of reproductive or sexual preference.

The success of an organism in leaving a numerous posterity is not measured only by the number of its surviving offspring, but also by the quality or probable success of these offspring. It is therefore a matter of importance which particular individual of those available is to be their other parent. With the higher animals means of discrimination exist in the inspection of the possible mate, for in large groups the sense organs are certainly sufficiently well developed to discriminate individual differences. It is possible therefore that the emotional reactions aroused by different individuals of the opposite sex will, as in man, be not all alike, and at the least that individuals of either sex will be less easily induced to pair with some partners than with others. With plants an analogous means of discrimination seems to exist in the differential growth rate of different kinds of pollen in penetrating the same style.

An excellent summary of recently established facts in this field has been given by D. F. Jones (*Selective Fertilization*, University of Chicago, 1928). Cases are known in maize in which discrimination is exercised against pollen bearing certain deleterious mutant factors, and in one case in *Oenothera* against ovules bearing a certain lethal factor. In these reactions both the genotype of the mother plant and that of the pollen are exposed to selection, and it is this that serves to explain the remarkable fact established by Jones' own observations with maize, that pollen applied in mixtures is on the whole less effective the greater the genetic diversity between the seed parents and the pollen parent. Such a generalized tendency towards homogamy, which is perhaps especially manifest in maize owing to the enormous number of recessive defects, which by continued cross pollination have accumulated in that plant, would, however, be far less effective in promoting the fission of species than would the selection of discriminative tendencies specially directed towards that end, such as must occur, as will be explained more fully below, in a group constantly invaded by the diffusion of unfavourable genes.

In general the conditions upon which discrimination, when possible, can usefully be exercised seem to be (i) that the acceptance of one mate precludes the effective acceptance of alternative mates, and (ii) that the rejection of an offer will be followed by other offers, either certainly, or with such high probability, that the risk of their non-occurrence shall be smaller than the probable advantage to be gained by the choice of a mate. The first condition is satisfied by the females of most species, and in a considerable number of cases by the males also. In other cases, while it would be a serious error for the male to pursue an already fertilized female, it would seem that any opportunity of effective mating could be taken with advantage. The second condition is most evidently satisfied when members of the selected sex are in a considerable majority at the time of mating.

The grossest blunder in sexual preference, which we can conceive of an animal making, would be to mate with a species different from its own and with which the hybrids are either infertile or, through the mixture of instincts and other attributes appropriate to different courses of life, at so serious a disadvantage as to leave no descendants. In the higher animals both sexes seem to be congenitally adapted to avoid this blunder and from the comparative rarity of natural hybridization among plants, save in certain genera where specific distinctness may have broken down through maladaptation in this very respect, we may infer the normal prevalence of mechanisms effective in minimizing the probability of impregnation by foreign pollen. It is therefore to be inferred that in the higher animals the nervous system is congenitally so constructed, that the responses normal to an association with a mate of its own species are, in fact, usually inhibited by the differences which it observes in the appearance or behaviour of a member of another species. Exactly what differences in the sensory stimuli determine this difference in response it is of course impossible to say, but it is no conjecture that a discriminative mechanism exists, variations in which will be capable of giving rise to a similar discrimination within its own species, should such discrimination become at any time advantageous.

A typical situation in which such discrimination will possess a definite advantage to members of both sexes must arise whenever a species occupying a continuous range is in process of fission into two daughter species, differentially adapted to different parts of that range ; for in either of the extreme parts certain relatively disadvantageous characters will constantly appear in a certain fixed proportion of the individuals in each generation, by reason of the diffusion of the genes responsible for them from other parts of the range. The individuals so characterized will be definitely less well adapted to the situation in which they find themselves than their competitors ; and in so far as they are recognizably so, owing, for example, to differences in tint, their presence will give rise to a selective process favouring a sexual preference of the group in which they live. Individuals in each region most readily attracted to or excited by mates of the type there favoured, in contrast to possible mates of the opposite type, will, in fact, be the better represented in future generations, and both the discrimination and the preference will thereby be enhanced. It appears certainly possible that an evolution of sexual preference due to this cause would establish an effective isolation between two differentiated parts of a species, even when geographical and other factors were least favourable to such separation.

An example of a species in process of fission, in which sexual preference is evidently playing an important part, occurs in butterflies of the genus *Limenitis* (Basilarchia) in the Eastern United States. The researches of W. L. W. Field of Milton, Massachusetts, have shown that the Northern species (or more properly sub-species) *L. arthemis*, distinguished by a conspicuous white band on both fore and hind wings, occasionally interbreeds with the Southern sub-species *astyanax*, yielding hybrids of both sexes, distinguishable from both parental forms. These hybrids, though relatively rare, are found throughout the narrow zone, from the Atlantic coast north of Boston to the Mississippi River, on which the ranges of the two sub-species overlap. The interpretation of the data is facilitated by the circumstance that the conspicuous white band in *L. arthemis* is due to a single Mendelian factor, in which that form differs from *astyanax*, although this is evidently not the only factor in which the forms differ. The white band is incompletely recessive and hybrid females caught wild, and yielding 50 per cent *arthemis*-like young, have thus shown themselves to be fertile with the one parental form. The occurrence, moreover, in the southern border of their range, of banded butterflies, in other respects much resembling *astyanax*, is evidence that the hybrids will breed with each other and possibly also with the Southern parent, *L. astyanax*.

The same observation also shows that the Southern form has, since its separation, accumulated other factorial contrasts with the Northern form, in addition to that which determines the suppression of the white band.

A most important feature, the significance of which has been pointed out by the late Professor Poulton, is that the rarity of the heterozygotes, in the zone in which they occur, implies that the butterflies in this zone display a strong preference in mating, each for its own kind. In addition to the factorial differences which determine the minor features of the wing pattern, differences which determine the reactions of the nervous system must also have been accumulated in such a way as to lead *astyanax* to mate preferentially with *astyanax* and *arthemis* with *arthemis*. The selective advantage by which these modifications have been brought about is, in this case, clear, for at the low frequency at which the hybrid occurs the vast majority of its matings must be with the parental forms, and, since such matings would yield 50 per cent. heterozygotes, unless these latter were at a selective disadvantage, they would maintain their own numbers, and be further reinforced in each generation by such crossings as naturally occur between the parental forms. It is probable, in fact, that heterozygosis is here never maintained for more than a few generations, and that almost the entire supply of heterozygotes comes from stray matings of the parent sub-species, which are thus on the verge of a complete genealogical separation.

It may also be noted that the closely allied Viceroy *L. archippus*, the appearance of which has been much modified by its close mimicry of the Monarch or Milkweed butterfly, *Anosea plexippus*, has a range completely overlapping those of the two sub-species discussed above. In this case where geographical isolation is not available, the genitalia of the male, *L. archippus*, have been distinctly modified. It has been, however, possible to breed the female of this species with *L. arthemis*, producing males only, which could not, however, be induced to mate. Such males have also, though very rarely, been observed in Nature. Specialization of the male genitalia here appears as a later stage in the process of fission than preferential mating, and indeed such specialized modification would appear only to be advantageous when preferential mating, or other causes, had already induced an almost complete fission of the sub-specific groups.

Sexual selection

The theory put forward by Darwin to account for the evolution of secondary sexual characters involves two rather distinct principles. In one group of cases, common among mammals, the males, especially when polygamous, do battle for the possession of the females. That the selection of sires so established is competent to account for the evolution, both of special weapons such as antlers, and of great pugnacity in the breeding season, there are, I believe, few who doubt, especially since the investigation of the influence of the sex hormones has shown how genetic modifications of the whole species can be made to manifest themselves in one sex only, and has thereby removed the only difficulty which might have been felt with respect to Darwin's theory.

For the second class of cases, for which the amazing development of the plumage in male pheasants may be taken as typical, Darwin put forward the bold hypothesis that these extraordinary developments are due to the cumulative action of sexual preference exerted by the females at the time of mating. The two classes of cases were grouped together by Darwin as having in common the important element of competition, involving opportunities for mutual interference and obstruction, the competition being confined to members of a single sex. To some other naturalists the distinction between the two types has seemed more important than this common element, especially the fact that the second type of explanation involves the will or choice of the female. A. R. Wallace accepted without hesitation the influence of mutual combats of the males in the evolution of sex-limited weapons, but rejected altogether the element of female choice in the evolution of sex-limited ornaments.

It has been pointed out in Chapter II that a detailed knowledge of the action of Natural Selection would require an accurate evaluation of the rates of death and reproduction of the species at all ages, and of the effects of all the possible genetic substitutions upon these rates. The distinction between one kind of selection and another would seem to require information in one respect infinitely more detailed, for we should require to know not the gross rates of death and reproduction only, but the nature and frequency of all the bionomic situations in which these events occur. The classification of causes of death required by law is sufficiently complex, and would require very extensive medical knowledge if full justice were to be done to it in every case. Even qualified medical men, however, are not required to specify the sociological causes of birth. In pointing out the immense complexity of the problem of discriminating to which possible means of selection a known evolutionary change is to be ascribed, or of allotting to several different means their share in producing the effect, I should not like to be taken to be throwing doubt on the value of such distinctions as can be made among the different bionomic situations in which selection can be effected. The morphological phenomena may be so striking, the life-history and instincts may have been so fully studied in the native habitat, that a mind fully stored with all the analogies within its field of study may be led to perceive that one explanation only, out of those which are offered, carries with it a convincing weight of evidence. Every case must, I conceive, be so studied and judged upon by persons acquainted with the details of the case, and even so in the vast majority of cases the evidence will be too scanty to be decisive. It would accord ill with the scope of this book (and with the pretensions of its author) to attempt such a decision in any particular case. There does seem room, however, for a more accurate examination of the validity of the various types of argument which have been used, and which must be used if any interpretation at all is to be put upon the evidence, than seems hitherto to have been attempted.

It is certain that some will feel that such an abstract form of treatment does injury to the interest of the subject. On the other hand I am confident that many engaged in the actual work of observation and classification would welcome any serious attempt to establish impartial principles of interpretation. The need is greatest in a subject, in which generalizations embodying large numbers of observational facts are of such high value, that in controversy mere citations of fresh facts seem sometimes to be invested with a logical force, which they do not really possess; it is possible thus for even the fairest minded of men, when thoroughly convinced of the correctness of his own interpretation, in which conviction he may be fully justified, to use in its support arguments which, had he been in real doubt, he could scarcely have employed. To take but a single instance of a most innocent lapse of logic in discussions of sexual selection; it was pointed out by Wallace that very many species which are conspicuously or brilliantly coloured, and in which the females are coloured either exactly like the males, or, when differently coloured are equally conspicuous, either nest in concealed situations such as holes in the ground or in trees, or build a domed or covered nest so as completely to conceal the sitting bird. In this concealment Wallace perceived an explanation of the lack of protective coloration in the female. To the objection, which seems to have originated with the Duke of Argyll, that a large domed nest is more conspicuous to an enemy than a smaller open nest, Wallace replied that as a matter of fact they do protect from attack, for hawks or crows do not pluck such nests to pieces. Darwin, on the other hand, believed that there was much truth in the Duke of Argyll's remark, especially in respect to all tree-haunting carnivorous animals. It will be noticed that neither controversialist seems to perceive that the issue is not concerned with the advantages or disadvantages of covered nests, or that, however disadvantageous these nests may be supposed to be, they nevertheless do fulfil the conditions required by Wallace of precluding the selection during brooding of protective colours in the female, by the action of predators to which brooding females might otherwise have been visible.

A much more serious error, which has not been without echoes in biological opinion, was made by Wallace in arguing that the effect of selection in the adult is diminished by a large mortality at earlier stages (*Darwinism*, p. 296).

In butterflies the weeding out by natural selection takes place to an enormous extent in the egg, larva, and pupa states; and perhaps not more than one in a hundred of the eggs laid produces a perfect insect which lives to breed. Here, then, the impotence of female selection, if it exists, must be complete; for, unless the most brilliantly coloured males are those which produce the best protected eggs, larvae, and pupae, and unless the particular eggs, larvae, and pupae, which are able to survive, are those which produce the most brilliantly coloured butterflies, any choice the female might make must be completely swamped. If, on the other hand, there *is* this correlation between colour development and perfect adaptation at all stages, then this development will necessarily proceed by the agency of natural selection and the general laws which determine the production of colour and of ornamental appendages.

It should be observed that if one mature form has an advantage over another, represented by a greater expectation of offspring, this advantage is in no way diminished by the incidence of mortality in the immature stages of development, provided there is no association between mature and immature characters. The immature mortality might be a thousandfold greater, as indeed it is if we take account of the mortality of gametes, without exerting the slightest influence upon the efficacy of the selection of the mature form. Moreover, Wallace himself attached great importance to other selective effects exerted upon mature butterflies as is shown by his treatment of protective resemblance on page 207 of the same work. It cannot therefore have been the cogency of the argument he uses which determined Wallace's opinion, but rather the firmness of his conviction that the aesthetic faculties were a part of the 'spiritual nature' conferred upon mankind alone by a supernatural act, which supplies an explanation of the looseness of his argument.

The two fundamental conditions which must be fulfilled if an evolutionary change is to be ascribed to sexual selection are (i) the existence of sexual preference at least in one sex, and (ii) bionomic conditions in which such preference shall confer a reproductive advantage. In cases where the two conditions can be satisfied, the existence of special structures, which on morphological grounds may be judged to be efficacious as ornaments, but to serve no other useful purpose, combined with ecological evidence that the structures are at their fullest development in the mating season, and are then paraded conspicuously, provides evidence of the same kind as, in other cases, is deemed conclusive as to the evolutionary significance of bodily structures.

With respect to sexual preference, experimental evidence of its existence in animals other than man is, and perhaps always will be, meagre. The only point of value, which it would seem might be determined by observation of birds in captivity, is the extent to which different hen birds concur in their preferences among the cocks. Since, I suppose, only one choice could be confidently observed in each season, such a test could only be applied with sufficient numbers to polygamous birds; among these, however, it should be possible to demonstrate it with certainty, if an order of preference exists.

The strongest argument adduced by Darwin as to monogamous birds in wild state must certainly be given now a different interpretation. He gives numerous cases in which when one of a pair of birds is shot, its place is found almost immediately to be taken by another of the same sex, whether male or female. He concludes that, surprising as it may seem, many birds of both sexes remain unpaired, and also— and here only we part company with him—this, because they cannot find a mate to please them. If this were the true explanation it would indicate sexual preferences so powerful as to inhibit mating altogether in a considerable proportion of birds, and such intensity of preference could scarcely be maintained in a species unless the advantage to the prospects of the progeny due to the possibility of gaining a very superior mate were larger than the certain loss of an entire breeding season. As will be seen, it is difficult to assign in most cases a rational basis for so great an advantage. On the other hand the researches of H. E. Howard upon *Territory in Bird Life* provide a very simple and adequate explanation of the fact observed. On this view the birds which remain unmated do so because they are not in possession of a breeding territory where they can nest unmolested, but are ready to mate at once with a widow or widower left in possession of this coveted property. I do not know, however, how much evidence there is for asserting that it is always the widowed bird and a new mate, rather than a new pair, which is found in possession of the vacant territory. The adoption of existing young suggests that sometimes at least it is the former.

If instead of regarding the existence of sexual preference as a basic fact to be established only by direct observation, we consider that the tastes of organisms, like their organs and faculties, must be regarded as the products of evolutionary change, governed by the relative advantage which such tastes may confer, it appears, as has been shown in a previous section, that occasions may be not infrequent when a sexual preference of a particular kind may confer a selective advantage, and therefore become established in the species. Whenever appreciable differences exist in a species, which are in fact correlated with selective advantage, there will be a tendency to select also those individuals of the opposite sex which most clearly discriminate the difference to be observed, and which most decidedly prefer the more advantageous type. Sexual preference originating in this way may or may not confer any direct advantage upon the individuals selected, and so hasten the effect of the Natural Selection in progress. It may therefore be far more widespread than the occurrence of striking secondary sexual characters.

Certain especially remarkable consequences do follow if some sexual preferences of this kind, determined, for example, by a plumage character, are developed in a species in which the preferences of one sex, in particular the female, have a great influence on the number of offspring left by individual males. In such cases the modification of the plumage character in the cock proceeds under two selective influences (i) an initial advantage not due to sexual preference, which advantage may be quite inconsiderable in magnitude, and (ii) an additional advantage conferred by female preference, which will be proportional to the intensity of this preference. The intensity of preference will itself be increased by selection so long as the sons of hens exercising the preference most decidedly have any advantage over the sons of other hens, whether this be due to the first or to the second cause. The importance of this situation lies in the fact that the further development of the plumage character will still proceed, by reason of the advantage gained in sexual selection, even after it has passed the point in development at which its advantage in Natural Selection has ceased. The selective agencies other than sexual preference may be opposed to further development, and yet the further development will proceed, so long as the disadvantage is more than counterbalanced by the advantage in sexual selection. Moreover, as long as there is a net advantage in favour of further plumage development, there will also be a net advantage in favour of giving to it a more decided preference.

The two characteristics affected by such a process, namely plumage development in the male, and sexual preference for such developments in the female, must thus advance together, and so long as the process is unchecked by severe counterselection, will advance with ever-increasing speed. In the total absence of such checks, it is easy to see that the speed of development will be proportional to the development already attained, which will therefore increase with time exponentially, or in geometric progression. There is thus in any bionomic situation, in which sexual selection is capable of conferring a great reproductive advantage, as certainly occurs in some polymorphic birds, the potentiality of a runaway process, which, however small the beginnings from which it arose, must, unless checked, produce great effects, and in the later stages with great rapidity.

Such a process must soon run against some check. Two such are obvious. If carried far enough, it is evident that sufficiently severe counterselection in favour of less ornamented males will be encountered to balance the advantage of sexual preference; at this point both plumage elaboration and the increase in female preference will be brought to a standstill, and a condition of relative stability will be attained. It will be more effective still if the disadvantage to the males of their sexual ornaments so diminishes their numbers surviving to the breeding season, relative to the females, as to cut at the root of the process, by diminishing the reproductive advantage to be conferred by female preference. It is important to notice that the condition of relative stability brought about by these or other means, will be of far longer duration than the process in which the ornaments are evolved. In most existing species the runaway process must have been already checked, and we should expect that the more extraordinary developments of sexual plumage were not due like most characters to a long and even course of evolutionary progress, but to sudden spurts of change. The theory does not enable us to predict the outcome of such an episode, but points to a great advantage being conferred by sexual preference as its underlying condition.

Exactly in what way the males which most effectually attract the attention and interest of the females gain thereby a reproductive advantage is a much more difficult question, since polygamy is not nearly so widespread as sex-limited ornaments; the theory of sexual selection therefore requires that some reproductive advantage should be conferred also in certain monogamous birds. Darwin's theory on this point is exceedingly subtle. He supposes in effect that there is a positive correlation in the females between the earliness with which they are ready to breed, and the numbers of offspring they rear, variations in both these variates being associated, as Darwin suggests, with a higher nutritional condition. Whether this is so in fact it is difficult to say, but it should be noted that the dates of the breeding phenomena of a species could only be stabilized if birds congenitally prone to breed early did not for this reason produce more offspring. The correlation required by Darwin's theory must be due solely to non-hereditary causes, such as chance variations of nutrition might supply. Whether or not there is such a correlation, it would seem no easy matter to demonstrate.

The table shows hypothetical average numbers of offspring reared by females differing in two respects, (a) congenital tendency to breed early, (b) nutritional condition which favours both early breeding and number of offspring. The indices represent the relative numbers of females in each class, out of a total of 256. Each row refers to a group of females with the same congenital response to the stimuli initiating the breeding sequence, the latest breeders being in the top row, and shows the frequencies of five different nutritional conditions, with the average numbers of offspring reared. Each column refers to birds actually breeding at the same time. The numbers of offspring are adjusted to increase with the nutritional condition of the female in each row, and at the same time to give a small further advantage to those breeding at or near the mean or optimal breeding date as opposed to those breeding late or early in the season. The selective effect upon the cocks is shown in the lower margin of average offspring according to breeding date, those chosen by the hens actually ready to breed early rearing the larger families. The selective effect upon the hens is shown in the right-hand margin, there being a slight elimination of hens congenitally prone to breed too early or too late, but no tendency to accelerate or retard the breeding date of the whole species.

The following scheme is an attempt to represent Darwin's theory in quantitative terms.

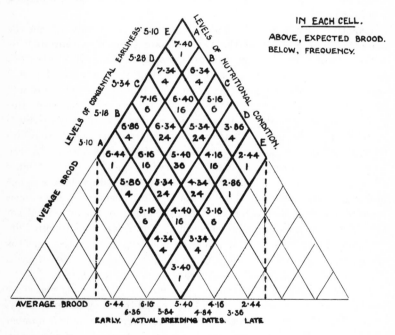

Schematic representation of Darwin's theory of sexual selection in monogamous birds, as interpreted by the author; showing the possibility of a selective advantage of males chosen by reason of superior adornment by early-breeding females, without any selective advantage of females congenitally prone to breed early.

There does seem also to be another advantage enjoyed by the **males mated earliest in any one district, and which therefore might** be conferred by sexual preference; namely, that due to mortality during the breeding season. The death rates of animals are often surprisingly high, and a death rate of only one per cent per week would give a considerable advantage to the earlier mated males, even if the chances of survival of his offspring were unfavourably affected by his death.

A second circumstance in which sexual preference must afford some reproductive advantage is in the remating of birds widowed during the breeding season; it appears certain that an abundance of unmated birds are usually at hand to take advantage of such a situation, and the choice among these gives to those preferred a reproductive advantage, equally in the case of both sexes. To judge, however, of the relative efficacy of the different possible situations in which sexual preference may confer a reproductive advantage, detailed ecological knowledge is required.

The possibility should perhaps be borne in mind in such studies that the most finely adorned males gain some reproductive advantage without the intervention of female preference, in a manner analogous to that in which advantage is conferred by special weapons. The establishment of territorial rights involves frequent disputes, but these are by no means all mortal combats; the most numerous, and from our point of view, therefore, the most important cases are those in which there is no fight at all, and in which the intruding male is so strongly impressed or intimidated by the appearance of his antagonist as not to risk the damage of a conflict. As a propagandist the cock behaves as though he knew that it was as advantageous to impress the males as the females of his species, and a sprightly bearing with fine feathers and triumphant song are quite as well adapted for war-propaganda as for courtship.

The selective action here considered combines the characteristics of the two classes to which Darwin applied the term sexual selection, namely the evolution of special weapons by combats between rival males, and the evolution of adornments which attract or excite the female. An appearance of strength and pugnacity is analogous to the possession of these qualities in producing the same effect; but the effect is produced in a different way, and in particular, as in the case of attractive ornaments, by the emotional reaction of other members of the species. It involves in fact closely similar mental problems to those raised by the existence of sexual preference. One difference should be noted; in the case of attractive ornaments the evolutionary effect upon the female is to fit her to appreciate more and more highly the display offered, while the evolutionary reaction of war paint upon those whom it is intended to impress should be to make them less and less receptive to all impressions save those arising from genuine prowess. Male ornaments acquired in this way might be striking, but could scarcely ever become extravagant.

Sex limitation of modifications

A difficulty which was regarded rather seriously during the development of the theory of sexual selection is implicit in the limitations of many of the structures ascribable to sex-limited selection, to the particular sex on which the selection acts. The difficulty lay in how far selection acting on only one sex ought to be expected to affect the characters of both sexes, and whether a mutation originally affecting the development of both sexes could be confined to one sex only, by counterselection on the other sex.

Of the large mutational changes chiefly available for genetic study the great majority manifest themselves equally in the two sexes; in an important minority the effect is either unequal in the two sexes or strictly limited to one sex. In birds and mammals a clearly understood mechanism of sexual differentiation lies in the internal secretions of the gonads, which are sexually differentiated, and possibly in a sexual differentiation of other internal secretions. It is a natural inference that a proportion of the mutations which occur affecting any given structure will be, from the first, sex limited in their appearance, and, if they produce their effect only in conjunction with the internal secretions of the sexual glands of one sex, their appearance will be delayed to the adult stage like the other signs of sexual maturity. This proportion may be as low as that observed in the genetic mutations, and indeed the only reason for thinking that it may be higher is that these somewhat violent changes may perhaps be expected to be produced by deviations occurring at an early stage of development, while the slighter changes to which progress by Natural Selection must chiefly be due may more frequently be initiated at later developmental stages.

Whatever the frequency, however, of sex limitation, it may fairly be inferred that selection acting upon one sex only, would, in the complete absence of counter-selection in the other sex, lead to an evolutionary modification not very unequal in the two sexes. Well-marked sexual differentiation must on this view be ascribed to a condition in which the selective agencies acting on the two sexes oppose each others influence. On the view that both sexes are in most species highly adapted to their place in nature, this situation will be readily brought about by the selection of modifications in one sex only, for these, in so far as they are not sex-limited, will be accompanied by changes in the opposite sex, which, on the assumption of high adaptation, will generally be disadvantageous and therefore opposed by selective agencies. Without the assumption of high adaptation, opposition between the actions of selection on the two sexes must be fortuitous and rare, and it is by no means clear how the widespread occurrence of sex-limited modifications can on such a view be explained.

Since the whole body of genetic evidence seems to favour, and even to require, the view that organisms are in general extremely closely adapted to their situations, we need only consider the consequences of this view. Selection applied to particular qualities in one sex only will then tend, in the first instance, to modify this sex slightly more than the other. The opposite sex will only be modified so far as to bring into play agencies exerting selection, in the opposite direction, and with equal intensity. The advantage of protective coloration, stressed by Wallace, is of obvious importance in this connexion. From this point, which must be reached relatively rapidly, onwards, the selective advantage of a mutation in respect of the selective activity under consideration, will not depend at all upon the average of its effects in the two sexes, but only upon the difference between these effects. If it enhances the sexual contrast and makes the two sexes less alike, it will be favoured by selection, and will have therefore a definite probability of contributing its quotum towards the building up of sexual differentiation. If on the contrary its effect would have been to render the sexes more alike, it will be rejected by the selection in progress. In this, the more prolonged evolutionary phase, it should be noted that any effect which the new mutations may have upon both sexes equally, or, in fact, the average of their effects upon the two sexes, will be immediately neutralized by a change of frequency in those factors which, without being sex limited, influence the development of the organ in question.

Besides the mutations, the effects of which are conditioned by the

sexual secretions, an important class of mutations are those which
influence the nature of these secretions themselves; for in the condition of sexually opposed selections, any modification of these
secretions, which, without impairing their normal action, enhances or
increases the range of their developmental effects will thus afford
a further means of increasing sexual differentiation. In this way it is
by no means a supposition to be excluded as impossible that a
character at first manifested equally by the two sexes should, by the
action of natural selection, later become sex-limited in its appearance.

Natural Selection and the sex-ratio

The problem of the influence of Natural Selection on the sex-ratio
may be most exactly examined by the aid of the concept of reproductive value developed in Chapter II. As is well known, Darwin expressly
reserved this problem for the future as being too intricate to admit
of any immediate solution. (*Descent of Man*, p. 399).

In no case, as far as we can see, would an inherited tendency to
produce both sexes in equal numbers or to produce one sex in excess,
be a direct advantage or disadvantage to certain individuals more than
to others; for instance, an individual with a tendency to produce more
males than females would not succeed better in the battle for life than
an individual with an opposite tendency; and therefore a tendency of
this kind could not be gained through natural selection. Nevertheless,
there are certain animals (for instance, fishes and cirripedes) in which
two or more males appear to be necessary for the fertilization of the
female; and the males accordingly largely preponderate, but it is by
no means obvious how this male-producing tendency could have been
acquired. I formerly thought that when a tendency to produce the two
sexes in equal numbers was advantageous to the species, it would follow
from natural selection, but I now see that the whole problem is so
intricate that it is safer to leave its solution for the future.

In organisms of all kinds the young are launched upon their careers
endowed with a certain amount of biological capital derived from
their parents. This varies enormously in amount in different species,
but, in all, there has been, before the offspring is able to lead an
independent existence, a certain expenditure of nutriment in addition,
almost universally, to some expenditure of time or activity, which
the parents are induced by their instincts to make for the advantage
of their young. Let us consider the reproductive value of these off-

spring at the moment when this parental expenditure on their behalf has just ceased. If we consider the aggregate of an entire generation of such offspring it is clear that the total reproductive value of the males in this group is exactly equal to the total value of all the females, because each sex must supply half the ancestry of all future generations of the species. From this it follows that the sex ratio will so adjust itself, under the influence of Natural Selection, that the total parental expenditure incurred in respect of children of each sex, shall be equal; for if this were not so and the total expenditure incurred in producing males, for instance, were less than the total expenditure incurred in producing females, then since the total reproductive value of the males is equal to that of the females, it would follow that those parents, the innate tendencies of which caused them to produce males in excess, would, for the same expenditure, produce a greater amount of reproductive value; and in consequence would be the progenitors of a larger fraction of future generations than would parents having a congenital bias towards the production of females. Selection would thus raise the sex-ratio until the expenditure upon males became equal to that upon females. If, for example, as in man, the males suffered a heavier mortality during the period of parental expenditure, this would cause them to be more expensive to produce, for, for every hundred males successfully produced expenditure has been incurred, not only for these during their whole period of dependance but for a certain number of others who have perished prematurely before incurring the full complement of expenditure. The average expenditure is therefore greater for each boy reared, but less for each boy born, than it is for girls at the corresponding stages, and we may therefore infer that the condition toward which Natural Selection will tend will be one in which boys are the more numerous at birth, but become less numerous, owing to their higher death-rate, before the end of the period of parental expenditure. The actual sex-ratio in man seems to fulfil these conditions somewhat closely, especially if we make allowance for the large recent diminution in the deaths of infants and children; and since this adjustment is brought about by a somewhat large inequality in the sex ratio at conception, for which no *a priori* reason can be given, it is difficult to avoid the conclusion that the sex-ratio has really been adjusted by these means.

The sex-ratio at the end of the period of expenditure thus depends upon differential mortality during that period, and if there are any such differences, upon the differential demands which the young of such species make during their period of dependency; it will not be influenced by differential mortality during a self-supporting period; the relative numbers of the sexes attaining maturity may thus be influenced without compensation, by differential mortality during the period intervening between the period of dependence and the attainment of maturity. Any great differential mortality in this period will, however, tend to be checked by Natural Selection, owing to the fact that the total reproductive value of either sex, being, during this period, equal to that of the other, whichever is the scarcer, will be the more valuable, and consequently a more intense selection will be exerted in favour of all modifications tending towards its preservation. The numbers attaining sexual maturity may thus become unequal if sexual differentiation in form or habits is for other reasons advantageous, but any great and persistent inequality between the sexes at maturity should be found to be accompanied by sexual differentiations, having a very decided bionomic value.

Summary

A consequence of sexual reproduction which seems to be of fundamental importance to evolutionary theory is that advantageous changes in different structural elements of the germ plasm can be taken advantage of independently; whereas with asexual organisms either the genetic uniformity of the whole group must be such that evolutionary progress is greatly retarded, or if there is considerable genetic diversity, many beneficial changes will be lost through occurring in individuals destined to leave no ultimate descendants in the species. In consequence an organism sexually reproduced can respond so much more rapidly to whatever selection is in action, that if placed in competition on equal terms with an asexual organism similar in all other respects, the latter would certainly be replaced by the former.

In order to take full advantage of the possible occurrence of advantageous mutations, mutation rates must be generally so low that in the great majority of loci the homologous genes throughout a single species are almost completely identical, and this is the condition which we appear to find in the higher organisms. With sexual reproduction species are not arbitrary taxonomic units such as they would be with asexual reproduction only, but are bound together by sharing a very complete community of ancestry, if we look back only a hundred generations. The bulk of intraspecific variance apart from the differences between geographic races, in which some degree of isolation has taken effect, must be ascribed to segregation during comparatively few generations in the immediate past.

Selection acting differently on different parts of a species, whether or not these parts are distinguished geographically, will induce distinctions between them in the frequency with which different genes or gene combinations occur, without necessarily impairing the unity of the species. An element of instability will, however, be introduced in such cases, by genetic modifications affecting the frequency of germinal interchange between the parts; and this, under sufficiently intense selection, will lead to the fission of species, even in the absence of geographical or other barriers to intercourse.

An important means of fission, particularly applicable to the higher animals, lies in the possibility of differential sexual response to differently characterized suitors. Circumstances favourable to the fission of species into parts adapted to different habitats will also be favourable to the development both of discrimination and of sexual preference.

The main postulate of Darwin's theory of sexual selection, namely the exercise of sexual preference, will thus tend to be satisfied by the effects of previous selection. We may infer that the rudiments of an aesthetic faculty so developed thus pervade entire classes, whether or not this faculty is in fact afforded opportunities of inducing evolutionary change. In species so situated that the reproductive success of one sex depends greatly upon winning the favour of the other, as appears evidently to be the case with many polygamous birds, sexual selection will itself act by increasing the intensity of the preference to which it is due, with the consequence that both the feature preferred and the intensity of preference will be augmented together with ever-increasing velocity, causing a great and rapid evolution of certain conspicuous characteristics, until the process can be arrested by the direct or indirect effects of Natural Selection.

Consideration of the mechanism of sex-limited hormones, by which the secondary sexual characteristics of mammals and birds are largely controlled, shows that sexual differentiation may be increased or diminished by the action of Natural Selection, either through the occurrence of mutations sex-limited in effect, or through a modification of the hormone mechanism. It is thus not impossible that a mutant form, at first manifested equally by both sexes, should later, under the action of selection, become confined to one sex only. The question of the ratio of the sexes at maturity has not the same importance for sexual selection as was formerly thought, at least in species in which the number of breeding pairs is limited by the allocation of territory. It is shown that the action of Natural Selection will tend to equalize the parental expenditure devoted to the production of the two sexes; at the same time an understanding of the situations created by territory will probably reveal more than one way in which sexual preference gives an effective advantage in reproduction.

VII

MIMICRY

The relation of mimicry theory to the parent theory of Natural Selection. Theories of Bates and Müller. Supposed statistical limitation of Müllerian theory. Observational basis of mimicry theory. The evolution of distastefulness. The theory of saltations. Stability of the gene-ratio. Summary.

If our object is to ascertain how living things have become what they are . . . , a solution can never be attained unless the details of the selective process are studied at least as fully and thoroughly as the material which is subjected to selection. POULTON, 1908.

THERE are three respects in which the theory of mimicry is of great importance to the student of Natural Selection. In the first place a great array of striking facts is by its means rendered intelligible under the heading of a single principle ; that principle being the selective action of a single definite factor, usually predatism, in an insect's environment, upon a single set of characters, coloration, patterning, attitude at rest, mode of flight, observable in the insect itself. In our ordinary state of ignorance as to the lives of wild creatures it is rare to be able to particularize, as can be done in this case, either the incidence of an environmental factor, or the peculiar benefits of an observed morphological or physiological characteristic.

Secondly, as has been mentioned on page 54, it is of special importance to be able to demonstrate, in any extensive group of cases, the adaptive significance of the characteristics of species. For while it may be true, as has been urged by Robson, that even a minute morphological examination of the differences distinguishing closely related species, is not usually capable of revealing to us their adaptive significance in relation to their natural habitats, there can be no doubt that the study of mimicry has shown in numerous instances the particular environmental factor to which are to be ascribed certain characteristics distinctive of nearly related species, sub-species, and even of local varieties.

Thirdly, it is a matter of historical interest that the theory of mimicry, as the greatest post-Darwinian application of Natural Selection, played an especially important part towards the end of the nineteenth and the beginning of the twentieth century (when the concentration of biological effort in the museums and laboratories was beginning to render the conclusions of the field naturalists of an earlier generation in part unintelligible), by calling constant attention to the importance of ecological observations for the interpretation of the material gathered in the great museums.

It will be clearly understood, from the argument of Chapter I, that in the opinion of the author the bearing of genetical discoveries, and in particular of the Mendelian scheme of inheritance, upon evolutionary theory, is quite other than, and indeed opposite to, that which the pioneers of Mendelism originally took it to be. These were already, at the time of the rediscovery of Mendel's work, in the full current of that movement of evolutionary thought, which in the nineties of the last century had set in in favour of discontinuous origin for specific forms. It was natural enough therefore that the discontinuous elements in Mendelism should, without sufficiently critical scrutiny, have been interpreted as affording decisive evidence in favour of this view. Nowhere perhaps in biology has this current of opinion introduced such serious discrepancies, as in the interpretation of mimetic resemblance.

Theories of Bates and Müller

The theory of Bates, put forward in 1861, implies no more than is readily understood from the term mimicry; namely that certain palatable forms, which Bates observed especially among butterflies, being preyed upon by insect-eating birds, are so placed that it is advantageous for them to be mistaken for other objects, especially less palatable forms, which they somewhat resemble; and that the selective advantage so conferred upon those individuals in whom the resemblance is most complete, has led these species to become more and more perfect mimics of other species inhabiting the same district, relatively immune to attack, and therefore appropriate models. This theory depends on the errors of the predators, but on errors of the senses only, not of the judgement. The choice of food made by the bird may be made by instinct, perfect, though latent, in the egg, or may be the result of experience. In the latter case no account is taken of the stages during which the experience is gathered, but only of the final stage when the education is complete and a correct judgement as to diet finally formed. The eye may still be deceived; it is the essence of Bates's theory that it should be exposed to deception; but errors of judgement have no place in the theory.

Whether it be supposed to be fixed by heredity, or to be open to
modification by experience, the judgement of the predator does impose
one limitation upon Bates's theory. If hereditary, it must have been
moulded by Natural Selection, as a system of food preferences advantageous to the species, and could not long continue to apply to
those particular situations in which the loss due to rejecting mimics
actually outweighed the advantage of rejecting the models. The same
consequence follows, only more immediately, if individual experience
is the basis of the judgement formed. It must be assumed, therefore,
that the mimics are sufficiently rare, or the models sufficiently
noxious, for it to be advantageous for the predator to reject the
models, together with a certain proportion of deceptive mimics, on
those occasions on which a discriminative rejection is practised; or
alternatively, to assume that the state of affairs observed is a transient
one, pending a more perfect adaptation of the predator's instincts. In
point of fact the conclusion was early drawn by Bates, and widely
accepted, that the mimic must be a comparatively rare species, gaining its advantage through resemblance to a highly protected and
abundant species inhabiting the same region.

A limitation of a similar kind, but of a somewhat different specifica-
tion, is imposed by the reaction of the model. The resemblance which
is favourable to the mimic will be for the same reason disadvantageous
to the model. An individual of the model species may suffer from
being mistaken for a mimic, and as the probability of this is least
when the resemblance is least, selection will tend to modify the model
so as to render it different from the mimic as conspicuously as possible.
This situation is not essentially different from that which gives rise
to warning colours generally, for to be recognized as unpalatable
is equivalent to avoiding confusion with palatable species. Close
Batesian mimicry can therefore only be established if the rate of
modification of the model has been less than that of the mimic, and
this may be taken, in terms of the predominant factors of the situation, to imply that the selective advantage conferred on the individual
mimic, exceeds the selective disadvantage suffered by the individual
model. Disparity in numbers is as useful in ensuring the fulfilment
of this condition as it is in satisfying that imposed by the feeding
proclivities of the predator.

The observation, which was familiar to Bates, of the close superficial resemblance between very abundant, and apparently equally

protected, species, led Müller eighteen years later to formulate what
has been called the Müllerian, as opposed to the Batesian, theory of
mimicry, a term which it is still convenient to apply, although, as
Professor Poulton has pointed out, the term mimicry should in
strictness be confined to the theory of Bates. Müller's theory involves,
more than does that of Bates, a consideration of the ecological
situation in which the destruction of butterflies by birds actually
occurs. He points out that young birds, at least, do in fact learn
much by experience, and that during this process of self education
in what is and what is not good to eat, the total destruction suffered
by two unpalatable species will be diminished and ultimately halved,
if they come gradually to resemble one another so closely that the
lesson of avoidance learnt from the one will be equally applicable to
the other. An extension of the notion of education in this argument
was pointed out in 1915 by C. F. M. Swynnerton, who, on the basis
of very detailed observations, together with extensive experiments,
on the preferences of insect-eating birds, considers that, owing to a
partial failure of visual memory, the process of education is continued
throughout life by encountering, without necessarily attacking, the
aposematic types; and that the greater the numbers showing a given
warning colour, the more frequently is the memory reinforced, and,
in consequence, actually fewer mistaken attacks are made.

The Batesian mimic gains its advantage at the expense of the
predator which it deceives, and of the model whose life it endangers.
In the Müllerian system both species alike are mimic and model, each
reaps an advantage of the same kind, and both co-operate to confer
an advantage upon the predator by simplifying its education. The
predator which requires to frustrate the wiles of a Batesian mimic
should develop a keen and sceptical discrimination; while he will best
take advantage of the Müllerian situation by generalization, and
reasoning from analogy. The predator will tend to co-operate with
both kinds of prey to establish the Müllerian relationship, which is
to the advantage of all three, while model and predator will both
tend to diminish the advantage of the Batesian mimic, the first by
directly diminishing the resemblance, the other by modification of
feeding habits and by increasingly keen discrimination. The last
tendency, while diminishing the advantage of resemblance of any
given degree, will, however, tend to increase the degree of resemblance
actually attained.

Supposed statistical limitation of Müllerian theory

It has been seen that the reactions of model and predator impose a statistical limitation upon the theory of Bates, of which the principal observable factor is the relative abundance of the species. It has been supposed, and the supposition has been frequently repeated, that a similar limitation inheres in the theory of Müller. In 1908 G. A. K. Marshall suggested that, for arithmetical reasons, of two equally unpalatable species inhabiting the same region the less numerous will tend to resemble the more numerous, while the more numerous will not reciprocate this tendency. Marshall does not suggest that the more numerous will tend to decrease the resemblance. The general purport of his paper is to emphasize the Batesian as opposed to the Müllerian factor in resemblance, and that principally for reasons other than that under discussion. The frequent repetition of his statistical argument gives this point a special importance, and it will be examined here since it is eminently of the type to which mathematical reasoning should be able to supply a decisive answer.

When no question of degree is introduced into the discussion nothing is clearer than the distinction between the Batesian and the Müllerian factors. If, however, we take into consideration that butterflies may exist in all degrees of palatability, and that avoidance or acceptance by the predator must depend greatly upon its appetite, there is some danger that the distinctness of the evolutionary tendencies pointed out by these two authors may be lost in the complexity of the actual biological facts. This is necessarily the case in the discussion of separate aspects of Natural Selection. An evolutionary tendency is perceived intuitively, and expressed in terms which simplify, and therefore necessarily falsify, the actual biological facts. The only reality which stands behind such abstract theories is, in each case, the aggregate of all the incidents of a particular kind, which can occur from moment to moment to members of a species in the course of their life-histories.

In order to show how these incidents may be specified let us imagine three species, occupying the same region, of which A is highly unpalatable, B less so, while C is free from objectionable qualities. Then all possible situations may be exhaustively classified as follows (I, II, III, IV), in which, however, the alternative reasons given for the occurrence of any situation (*a, b, c*), are by no means exhaustive, but might conceivably be much elaborated by detailed biological observations.

I. A, B, and C all liable to attack.

II. B and C liable to attack, but not A.

(*a*) A bird which, either by inheritance or sufficient experience, prefers B to A is ready to attack an object recognized as B and not to attack an object recognized as A.

(*b*) A bird has attacked A and found it to be unpalatable, but has not yet had sufficiently impressed upon its mind the unpalatable qualities of B.

III. A and C liable to attack, but not B.

(*c*) A bird has attacked B and found it to be unpalatable, but has not yet had sufficiently impressed upon its mind the unpalatable qualities of A.

IV. C only liable to attack.

In situation I, mimetic resemblance is without effect, while in situation IV, C would gain by being mistaken for A or B, and A or B would lose by being mistaken for C. C might, therefore, if the former effect were to exceed the latter, become a Batesian mimic of A or B, while, on the contrary, A and B would gain by emphasizing their distinctive colouring, if so they could diminish the danger of confusion with C. Similarly, in situation II, A loses by being mistaken for B, and B gains by being mistaken for A; while in situation III the reverse is the case; both A and B will seem, therefore, to be acted upon by opposing tendencies, one tending towards similarity, and the other towards dissimilarity. It is only when the situations are analysed into their suggested causes that it is possible to indicate the resultant effect.

For this purpose we distinguish the 'Batesian situation' II(a), from the 'Müllerian situations' II(*b*) and III(*c*), recognizing that this classification need not be exhaustive. It is then seen that Batesian situations are to be distinguished by (i) depending upon differences of palatability and (ii) producing a 'Batesian tendency' for B to approach A and for A to recede from B. While the Müllerian situations (*b*) and (*c*) do not depend upon any difference of palatability, but are taken to occur whenever both species are, on occasion, deemed inferior to some alternative food.

In the occurrence of situation II in which a bird will attack one form, but will not attack the other, the difference between the action of the Müllerian and the Batesian factor could only be discerned by an observer who knew whether the bird's judgement, admittedly right, was founded upon sufficient or insufficient experience. If the experience is sufficient the bird would never prefer A to B (situation III), consequently he would create only Batesian situations; while if his experience is insufficient we must admit that it might have led him to create situation III, and that even when he creates situation II, that situation is none the less Müllerian. As a Batesian agent a bird is always right in his judgement; as a Müllerian agent he may be right, but in view of his inexperience, he has no right to be right. If the bird, as is usually the case with man, has only partially satisfactory grounds for his judgement, he may in the same act be both a Batesian and a Müllerian agent. It is only in the statistical aggregates formed by all occasions of the two appropriate kinds that the two evolutionary tendencies are completely distinct. Further study and thought devoted to the habits of wild animals may thus enable other tendencies to be recognized.

It is not, however, obvious from the analysis developed above that the net effect of (b) and (c) will be to cause a mutual, though possibly unequal, approach between the two species, such as Müller inferred. It is Marshall's contention that when the unpalatability is equal, the less numerous species will be attracted by the greater, but the greater will not be attracted by the less. Marshall does not fail to draw from this conclusion a very important consequence, for, as he points out, his premises lead to the inevitable result that, when a mimetic similarity is once effected, the larger species will have gained the smaller share, but still a share, of advantage from the association, and one might be inclined to argue from this that the larger species also will be led to approach this more advantageous condition. The far-reaching conclusion is drawn that such an argument is not valid, unless a continuous path from the first state to the second can be shown to exist, such that the advantage increases for each step along the path. Such a restriction, if necessary, would throw upon the selectionist an onus of detailed demonstration, which his opponent might increase indefinitely by challenging the details with increasing minuteness. Even if the case of Müllerian mimicry were not in itself of sufficient importance, it would be essential to examine

in some detail any particular case in which the argument from ultimate advantage is believed to lead to an erroneous conclusion.

Marshall's argument is essentially as follows: if A and B are two
equally distasteful species, of which B is the less numerous, then, in
the absence of mistakes due to the resemblance of the two species, the
young birds will take a proportionately heavier toll of B than of A,
before they have all learnt their lesson; consequently any mutant of
A which resembles B will suffer more than the non-mutant type, and
in consequence will be eliminated. It will be seen that the mutant is
supposed to lose the whole of the advantage of the warning colour A,
and in return to receive only the less advantage of the warning colour
B, and this argument is indeed conclusive in showing that a mutation,
which leaps clear outside the protective influence of its type, will
suffer heavily for its rashness, even if, miraculously enough, its leap
lands it in the heart of the protective influence of a less numerous
aposeme. But what of a less violent mutation? Is it possible to gain
some of the advantage of resembling B without losing the whole of
the advantage of resembling A? Is it even possible that increased
shelter from aposeme B will more than counterbalance the loss from
decreased shelter from aposeme A? In his answer to Marshall's
argument Dixey puts forward a directly opposite supposition, namely
that a mutant of appearance intermediate between A and B, would
gain the full advantage of both resemblances. In fact, whereas
Marshall assumes that the whole of the advantage of resembling A
is lost before any of the advantage of resembling B is gained, Dixey
assumes, on the contrary, that the whole of the advantage of resembling B may be gained before any advantage of resembling A is
lost. Both are clearly extreme assumptions; neither can be true
generally, and since the two assumptions lead to opposite conclusions
it would seem, so far as these arguments carry us, that we are faced
with a balance of forces of unknown magnitude, and can neither
assert that the Müllerian principle will work, nor that it will fail.

There remains the argument upon which Müller relied, that the
final condition of close resemblance being beneficial to both species,
both will therefore tend to approach this advantageous condition.
Marshall challenges the legitimacy of this argument, his reason being
the powerful one that he has disproved the conclusion in a particular
instance; as we have seen, however, Marshall's argument in the
chosen instance is indecisive, and the general argument from the

advantage of the final state is in a position to reassert its claims. If it is conclusive, however, it should be possible to devise a form of argument which shall show unequivocally, on the agreed postulates, that the admitted Müllerian situation will in fact produce a Müllerian evolutionary tendency affecting both species concerned.

Such an argument may, I suggest, be constructed by comparing the fate of any deviation from the type A, not with the average of that type, but with an equally conspicuous but opposite deviation. It will be admitted that variations of the species A, whether due to mutation or to Mendelian recombination, will be equally frequent in the direction of B as in the opposite direction; we may, therefore, without error, consider such variations to occur in pairs comprising variations of equal magnitude, but in opposite directions.

Since they are of equal magnitude they will lose (if anything) equally by failing to be recognized as typically A, but if either, or both, are ever mistaken for species B, the greater benefit will certainly be reaped by the variation in the direction of B. Since the whole species may be regarded as made up of such pairs of variations, and since in every pair selection favours the one more like B, if either is favoured, the net resultant must be a modification in the direction of species B.

It will be seen that the condition for the existence of a mimetic tendency is that in a certain proportion of the situations in which A is liable to, but B is immune against attack, members of species A should, through their similarity to B, actually escape attack. This is somewhat different from the condition arrived at by Professor H. H. Turner, in his appendix to Poulton's *Memoir on Mimicry in the Butterflies of Fiji*, 1924, who speaks of an actual overlap of the variations of the two species as the condition for the efficacy of Müller's statistical argument. The possibility of error on the part of the predator seems an essential feature in mimicry theory, and allowance can be made for it in Turner's treatment, provided we interpret his distribution curves as referring, not to the objective variability of the species, but to the (probably much greater) variability of the predator's subjective impressions, influenced as these must often be by inattention or haste, and by deceptive or insufficient illumination— in fact, by whatever circumstances conduce to error, human or avian. It is rather remarkable that, on a subject so remote from direct evidence as the subjective impressions of birds, we should possess three good reasons for assuming an approximately normal distribution: (*a*) that the reasons for which this distribution is chosen as the 'normal law of errors' can scarcely be confined to mankind, (*b*) that the objective variability of a measurable character due either to Mendelian segregation, or to environmental fluctuations, is usually closely normal, and (*c*) that the resultant compounded of two independent distributions is necessarily more normal than one, and possibly than both of its components.

The argument developed above may assuredly be refuted by disproving any of the biological factors assumed in the discussion. If it were proved that situations never in fact arise in which a member of A would survive if mistaken for B, but would perish if not so mistaken, that no predator learns by experience or is ever influenced by mimetic resemblances, or that such variations of A as do favour the resemblance are not heritable, then the Müllerian theory of mimicry would fail as an explanation of the resemblances observed. The sole point established by the above reasoning is that if these biological factors are admitted the resulting evolutionary tendency cannot be confined to the less numerous of two species. The efficiency of Müllerian selection will doubtless be greater (*ceteris paribus*) with the smaller species, but the supposed statistical objection to the Müllerian attraction of a larger species (or group) by a smaller is wholly fictitious.

Observational basis of mimicry theory

Though it would be beyond the scope of this book to attempt even a general survey of the biological facts connected with mimicry; it is necessary to give a very concise summary of the kinds of observational data which suffice to put the theory beyond doubt as the only satisfying explanation.

The number of species involved, among the insects alone, is so great that in the majority of individual cases many classes of appropriate observations are lacking, and the probable explanation must therefore be judged from incomplete evidence, in the light of cases of which more is known.

The biological reasons upon which the conclusive cases are based have been developed with conspicuous success by Poulton, whose line of argument I can do little more than summarize, and to whose labours and inspiration the great mass of facts now accumulated is principally due.

(i) Mimetic resemblances bear the hallmark of *adaptation* in the multiplicity of the simultaneous modifications to which they are due. This feature, as was pointed out in Chapter II, cannot be simply exemplified, since the essence of adaptation lies in its complexity. Mimetic resemblances involve colour, pattern, form, posture, movement and sometimes also sound ; and many of these items, if analysed, are themselves highly complex. Natural Selection is the only means known to biology by which *complex adaptations* of structure to function can be brought about.

(ii) In addition to showing adaptation, mimetic resemblances manifest a further characteristic of Natural Selection, in the variety of methods by which the same end is attained. This characteristic, analogous to opportunism in human devices, seems to deserve more attention than it has generally received. For the mechanisms of living bodies seem to be built up far less than might be expected by a human inventor, on simple and effective mechanical, physical, or chemical principles. On the contrary every property of the behaviour of matter, however odd and extraneous it may seem, seems to have been pounced upon as soon as it happened to produce a desirable effect. In the case of mimicry Poulton mentions four distinct means used by different species of imitating the superficial appearance of an ant, and in a single mimicry group finds five different methods of giving an appearance of transparency to the wings of moths and butterflies.

In addition to these two features characteristic of Natural Selection generally, three classes of facts are available to establish the particular requirements of mimicry theory.

(iii) Mimetic resemblances in general are not to be explained by systematic affinity. Striking cases occur between different classes of animals. Wheeler describes a case of tactile mimicry in which a mite imposes upon the instincts of certain blind ants, by mimicking with its anterior legs those movements of the antennae, by means of which ants obtain food one from the other. Obviously the more closely allied are the organisms which show resemblance, the more frequently are homologous parts utilized in its elaboration, and the more care is needed to demonstrate that a superficial resemblance has been imposed upon or has prevented an initial divergence in appearance.

(iv) Mimetic resemblances are not accompanied by such additional similarities as do not aid in the production, or strengthening, of a superficial likeness.

PLATE II

FIGS. 1–3. The head of the abundant East African Acraeine butterfly *Acraea zetes acara*, Hew., as seen from the front (1), from above (2) and the side (3), showing that the palpi, which are inconspicuous in most butterflies, are a prominent feature with their orange colour displayed against the black background.

FIGS. 1A–3A. Similar aspects of the head of the Nymphaline butterfly *Pseudacraea boisduvali trimenii*, Butler, a mimic of *A. z. acara* and found in the same part of Africa. It is evident that the resemblance here extends to the exceptionally emphasized feature, as was observed in the American examples shown in Figs. 6 and 7 of Plate I.

FIG. 4. *Danaida tytia*, Gray, a conspicuous Oriental Danaine butterfly taken with its mimic (Fig. 5) in the Darjiling district.

FIG. 5. *Papilio agestor*, Gray, a swallowtail butterfly mimicking the pattern of *tytia*.

FIG. 6. *Neptis imitans*, Oberth., a Nymphaline butterfly from S. W. China, mimicking the geographical form of *D. tytia* which is found in the same area.

Thus these two butterflies of widely separated groups both mimic this peculiar Danaine pattern.

The butterflies and moths here represented illustrate by single examples the widespread mimicry of the chief distasteful families in the tropics—on Plate I the Ithomiinae (Fig. 6) and Heliconinae (Fig. 4) of the New World; on Plate II, the Danainae (Fig. 5), and Acraeinae (Figs. 1–3) of the Old.

PLATE II. MODELS AND MIMICS IN AFRICAN (*Figs.* 1–3*a*) **AND**
ORIENTAL (*Figs.* 4–6) INSECTS

(v) Mimetic resemblances occur between species inhabiting the same region, appearing at the same seasons and hours, and having frequently been captured flying together.

It should be added that it is not necessary, in order to establish the mimetic character of a resemblance, that all of these five classes of evidence should be available. Evidence under class (i) alone may be sufficiently strong to exclude all reasonable doubt. Although that under (iii), (iv), and (v) is of great value as independent corroboration, no one item of it is necessarily present in all mimetic examples. The importance of the facts grouped under (ii) lies not in their value as evidence for individual cases, but in the comparative and systematic survey of the phenomenon of mimicry as a whole.

Plates I and II, from examples chosen by Professor Poulton, will serve to exemplify the kinds of superficial similarities, which form the observational basis of the theory. While these are sufficient to illustrate some characteristics of the evidence, familiarity with the living animal and its ecological associates is of special importance in these studies.

To distinguish whether an observed resemblance is to be ascribed to the Batesian or to the Müllerian evolutionary tendencies, or to both acting simultaneously, or, as in the case of the ant parasites mentioned above, to some distinct type of selective action, appears to be a much more difficult matter. Although Müllerian selection, unlike that of Bates, causes a mutual approach of the two species, we can scarcely expect that their modifications will be sufficiently nearly equal, save in a small minority of cases, for both to be made apparent by comparison with related species. In the majority of cases modification will be manifest on phylogenetic grounds only in those species which have become most rapidly modified. The criterion of palatability seems even more difficult to apply; for though the principle of Bates is excluded when two species are actually equally acceptable or unacceptable, to demonstrate such equality with sufficient precision to exclude differential predatism, and to demonstrate it with respect to the effectual predatory population, would seem to require both natural knowledge and experimental refinement which we do not at present possess. It is perhaps from the lack of more clearly decisive means of discrimination, that reliance has been placed upon observations of relative abundance, which, though clearly relevant to the problem, are not in themselves sufficient to supply a final test. Batesian mimicry by a more numerous of a less numerous form, cannot be excluded as impossible on purely theoretical grounds; for if the model were extremely noxious or the mimic a not particularly valuable source of food, the motive for avoidance may be but little diminished by the increase of the mimic. Moreover it is not so much abundance relative to the entomological collector, as abundance relative to selective agents of unknown species, and whose habits and times of feeding are therefore also unknown, which has to be considered, when this argument is used to exclude the Batesian principle. Equally, demonstrable differences in palatability do not, as has been seen, serve to exclude the Müllerian factor, the potency of which appears to be manifest when similar warning patterns are adopted by a group of several different and remotely related species in the same district.

A distinction between the kind of resemblance attained by the Batesian and Müllerian factors might be theoretically inferred in cases in which the group of predators is confronted with only one or the other type of resemblance. Since the Batesian resemblance is deceptive, it should extend to every observable character, at least in so far that no conspicuous differences remain by which the mimic might be distinguished from the model. This is not a necessity for the warning colours developed by Müllerian selection; for these it might suffice that a single conspicuous character should induce the predator to classify the object viewed as unpalatable. Thus conspicuously different insects may enjoy the advantages of Müllerian selection provided they display in common any one conspicuous feature, whereas Batesian mimics should show at least some resemblance to their models in all features.

The evolution of distastefulness

An important question raised by both the Batesian and the Müllerian theories of mimicry concerns the process by which nauseous flavours, as a means of defence, have been evolved. Most other means of defence such as stings, or disagreeable secretions and odours, are explicable by increasing the chance of life of the individuals in which they are best developed, or of the social community to which they belong. With distastefulness, however, although it is obviously capable of giving protection to the species as a whole, through its effect upon the instinctive or acquired responses of predators, yet since any individual tasted would seem almost bound to perish, it is difficult to perceive how individual increments of the distasteful quality, beyond the average level of the species, could confer any individual advantage.

The gregarious habit of certain larvae supplies a possible solution of the problem, if we are willing to accept the view that the distasteful quality of the imago, which warning colours are so well adapted to advertise, is itself merely a by-product due to the persistence of nauseous substances acquired through the protection afforded to the larva. For, although with the adult insect the effect of increased distastefulness upon the actions of the predator will be merely to make that individual predator avoid all members of the persecuted species, and so, unless the individual attacked possibly survives, to confer no advantage upon its genotype, with gregarious larvae the effect will certainly be to give the increased protection especially to one particular group of larvae, probably brothers and sisters of the individual attacked. The selective potency of the avoidance of brothers will of course be only half as great as if the individual itself were protected; against this is to be set the fact that it applies to the whole of a possibly numerous brood. There is thus no doubt of the real efficacy of this form of selection, though it may well be doubted if all cases of insect distastefulness can be explained by the same principle.

Professor Poulton has informed me that distasteful and warningly coloured insects, even butterflies, have such tough and flexible bodies that they can survive experimental tasting without serious injury. Without the weight of his authority I should not have dared to suppose that distasteful flavours in the body fluids could have been evolved by the differential survival of individuals in such an ordeal. The effect of selection on gregarious larvae, while not excluding individual selection of the imago, provides an alternative which will certainly be effective in a usefully large class of cases.

The institution of well defined feeding territories among many birds in the breeding season makes it possible to extend the effect produced on gregarious larvae to other cases in which the larvae, while not gregarious in the sense of swarming on the same plant, are yet distributed in an area which ordinarily falls within the feeding territory of a single pair. The selective effect in such cases will be diluted in so far as larvae of other broods may fall within the same territory, and will share in the advantage or disadvantage occasioned by the high or low development of distastefulness of any larvae tasted within this area. Such dilution will certainly make slower the evolution of distastefulness, its speed, if we oversimplify the problem by supposing that a number of broods chosen at random share a common reputation, being inversely proportional to the number of broods; dilution of a selective effect, should, however, not be confused with a counteracting tendency, which might bring progress to a standstill. The laying habits of several moths with conspicuous, though scattered, larvae deserve attention in this respect.

The view that nauseous flavours have generally been acquired by the effects of selection acting upon related larvae living in propinquity, implies that gregariousness, or equivalent habits, were formerly used by species which are now distasteful, though it does not imply that species with distasteful and even conspicuous larvae should necessarily have retained the gregarious habit; for the advantages of this habit, among which we may surmise (i) the reduced exposure of the female during oviposition, and (ii) in the case of distasteful and conspicuous larvae the advantage of increased protection from predators, will not always counterbalance the disadvantage sometimes entailed by a depletion of the food-supply. We may, however, fairly infer that if gregarious or equivalent habits are a necessary condition for the development of protective flavours, and if the possession of such flavours gives an added advantage, both to the gregarious habit, and to conspicuous coloration, then some association should be observable in nature between the conspicuousness of the larvae, and the habit of laying the whole brood either in great numbers on a single plant, or on neighbouring plants. The subject is one on which a comprehensive summary of the biological facts is much to be desired. I am indebted to Professor Poulton and Dr A. D. Imms for the following instances.

The Buff-tip moth *Zygaera bucephala* is distasteful alike as larva, pupa, and imago. The caterpillar is conspicuously yellow and black, and strikingly gregarious, the pupa is brown and buried, though equally distasteful, and the perfect moth is a beautiful example of cryptic coloration. Since the distastefulness is thus only advertised in the larval stage it may well have been developed by selection in this stage only, a process certainly facilitated by its gregarious habits. The Gothic moth, *Mania typica* is also unpalatable at all stages; in this case the larvae are not conspicuous, but feed when small in companies on the under sides of leaves.

The hairy rather brightly coloured caterpillars of *Onethocampa processionea* live in communal webs. When they have eaten the foliage enclosed in the web they march out in large 'processions', and seek out a site for a new web. When older they sally out at dusk and break into small parties to feed, returning to the communal web before daylight; they seem instinctively to avoid solitude as if it were a danger. These caterpillars exhibit a number of features, the web, the avoidance of daylight, and the fear of solitude, each of which may be interpreted as affording some measure of protection against different predators.

Conspicuous larvae of butterflies and moths are generally gregarious though not always so. An apparent exception is afforded by the Magpie moth *Abraxas grossulariata* of which the larvae feed separately; the eggs, however, are laid in groups, and the larvae are numerous on the same bush, or on adjacent bushes. *Anosia plexippus* however, scatters her eggs, although she has solitary, inedible, conspicuous, larvae. This is probably true of many Danaines.

True gregariousness seems to be characteristic of distasteful species which fly over considerable distances, while those with sluggish females may have scattered, though not widely scattered larvae.

An instructive case is afforded by the Peacock and Tortoise-shell butterflies, *Vanessa*, and related genera. These are swift fliers, and their larvae might, therefore, be widely scattered. In the true genus *Vanessa* the larvae are conspicuous, and the eggs are laid in batches, whereas in the Painted Lady *Cynthia cardui*, and the Red Admiral *Pyrameis atalanta*, formerly placed in *Vanessa*, the larvae are partly concealed and the eggs are laid singly.

Larvae of many Chrysomelid beetles are gregarious, and some of these at least are notably distasteful.

Of Saw-fly larvae Cameron says those 'which give out secretions or fetid odours are gregarious, several feeding on the same leaf, often ranged in a row with their bodies stuck out in the air. They have nearly always bright colours'. The advantage of gregariousness in strengthening these means of defence is obvious; they may be regarded as developments of the passive defence afforded by nauseous flavours, and it is in the initial stages of this development that the gregarious habit seems to supply the condition for effective selection.

In general it may be said that the observational facts in this field are consistent with, and lend some support to the view, that, whereas offensive flavours supply the condition for the evolutionary development of warning colours, and presumably also of warning odours, the condition for the evolutionary development of offensive flavours, which may equally characterize all stages of the life history, is to be found in the gregariousness or propinquity of larvae of the same brood, and therefore of somewhat highly correlated genetic constitution. A much wider knowledge of the observational facts than the author possesses would, however, be needed, before it could be asserted that no alternative view would suffice to explain the evolution of distastefulness, as a means of defence.

The principle deduced in this section of the selective advantage shared by a group of relatives, owing to the individual qualities of one of the group, who enjoys no personal selective advantage, is analogous to the situation which arises in human communities, in the tribal state of organization, in the selection of the group of qualities which may be summed up as heroism. The ideal of heroism has been developed among such peoples considerably beyond the optimum of personal advantage, and its evolution is only to be explained, in terms of known causes, by the advantage which it confers, by repute and prestige, upon the kindred of the hero. The human situation, which will be analysed in more detail in Chapter XI is, however, certainly complicated by sexual selection, of which there is no evidence in the case of the evolution of distasteful qualities in insects.

The theory of saltations

In his book on *Mimicry in Butterflies*, 1915, Punnett repeats Marshall's argument, and concludes without reservation that Müllerian mimicry of a less numerous by a more numerous species is excluded by it.* At first sight the argument appears irrelevant to Punnett's main contention of the inadequacy of Natural Selection to produce adaptations, for he evidently, unlike Marshall, would reject also both Batesian mimicry, and the Müllerian mimicry of the more numerous by the less numerous species. Nevertheless, it would not be altogether fair to regard Punnett's citation of Marshall's argument as a merely extraneous addition to his indictment, such as by arousing suspicion of error, though on an irrelevant issue, might serve to secure a verdict on the main count; on the contrary, Marshall's argument plays a

* Professor Punnett has since informed me that he does not reject Batesian mimicry.

small yet essential part in his destructive argument derived from mimicry rings. The case of two presumably palatable female types each quite unlike the corresponding males, which males are unlike each other, is chosen to illustrate this difficulty. The two females show an apparent mimetic resemblance to three other butterflies, two regarded as definitely unpalatable and the third as doubtfully so. Assuming that the non-mimetic males represent the former appearance of the two mimetic females, it is asked how the latter have come to resemble the distasteful members of the ring. Granted that these models might once have been not unlike in appearance to one of these males it can scarcely be assumed that they ever resembled both, either simultaneously or consecutively; but unless such a resemblance formerly existed a *gradual* mimetic evolution is precluded, and we should be forced to admit that the mimetic females arose as sports or saltations totally unlike their mothers.

It will be seen that for Punnett's argument on this important point, the gradual and mutual convergence of two or more different warning colours must be wholly excluded, for if the possibility of such a process is admitted, the difficulty of imagining a continuous sequence of changes entirely disappears, while on the contrary the assumption of discontinuity becomes a burden upon the theory, involving as it does the definite improbability of hitting off a good resemblance at one shot. Consequently Marshall's argument, which Punnett seems to have taken as reimposing all the limitations of the Batesian situation, plays an essential part in the argument in favour of saltations; so essential indeed that it seems impossible to repair the breach made by its removal.

The case for saltations as presented by Punnett was not entirely negative and destructive in character; it embodied one (then) recently discovered fact of considerable interest, namely that the differences between the three forms of the trimorphic female of *Papilio polytes* could be ascribed to two Mendelian factors, both limited in their obvious effects to the female sex, and one apparently necessary for the manifestation of the second.

This fact is of importance as indicating the mechanism by which a clear polymorphism is maintained; it shows that polymorphism in this case, and probably in similar cases, is dependent on one or more Mendelian factors the function of which is to switch on one or other of the possible alternatives, just as the more widespread dimorphism of sex is also dependent upon the Mendelian mechanism. In some groups, e. g. *Drosophila* and Man, a whole chromosome is utilized in the process of sex determination, in some fishes, on the contrary, crossing over has been found to occur between the 'sex-chromosomes' in the heterogametic sex, whether male as in *Lebistes*, or female as in *Platypoecilus*; in fishes it appears that we ought more properly to speak of the sex-gene rather than the sex-chromosome as the agent of sex determination. The passage from the one condition to the other, by the cessation of crossing over, presents no inherent difficulties, especially as Mendelian factors are known which expedite or inhibit crossing over. The reason for such a change is not so obvious, but since both systems are found still in use, it is probable that each has, upon particular conditions, its own advantages.

The core of Punnett's argument in favour of the production of mimetic forms by saltations lies in the Mendelian behaviour of the polymorphic females, for it is argued that these Mendelian factors must have arisen originally as mutations, and seeing that the different forms demonstrably differ by only single factor differences, these types must have sprung into existence each at a single leap. Convincing as this argument at first sight seems, we should, nevertheless, at once recognize our fallacy if we argued that because the sex difference in the fish *Lebistes* is apparently determined by a single factor, therefore a female fish of that genus, with the appropriate adaptations of her sex had arisen by a single saltation from a male of the same species! Or *vice versa*. In this case we are freed even from the necessity of rejecting the supposed saltation as improbable, for since the reproduction of the species requires the co-operation of both sexes, we may be certain that the origin of the sex factor antedated the evolution of separate sexes, and has presumably persisted, in its function of switch, unchanged during the whole course of the evolutionary development of these two types.

The example of sex emphasizes strongly the fact, which is becoming more and more appreciated as genetical research is more applied to complex and practical problems, that it is the function of a Mendelian factor to decide between two (or more) alternatives, but that these alternatives may each be modified in the course of evolutionary development, so that the morphological contrast determined by the factor at a late stage may be quite unlike that which it determined at its first appearance. The inference, therefore, that because a single factor determines the difference between a mimetic and a male-like form in *P. polytes*, therefore the mimetic form arose fully developed by a single mutation, is one that cannot fairly be drawn; it requires, in fact, the gratuitous assumption that no evolutionary change has taken place in either of the two alternative forms since the dimorphism was first established.

Certain genetical experiments have demonstrated that genetic changes of the kind here considered are compatible with a purely Mendelian scheme of inheritance. In rats, the hooded (black and white) pattern is a simple recessive to the 'self' or 'solid' coloration; the case is probably parallel to the 'Dutch' pattern in rabbits, and the 'recessive pied' in mice. In studying variations in the hooded pattern Castle found that by selection it was easy to obtain strains of hooded rats which were almost entirely black, and other strains almost entirely white, and equally, of course, a large number of stable patterns of an intermediate character. All these types of 'hooded' behaved, as before selection, as simply recessive to self-colour. Two possible explanations were put forward; the first possibility was that the modification produced by selection lay in the hooded gene, that, in fact, selection had sorted out from a large number of slightly differing allelomorphs, those favouring much or little pigmentation, and consequently that the surviving hooded genes were different from those prevalent before selection. The second possibility was that the hooded gene was invariable in character, but that the pigmented area depended also on the co-operation of other genes, so-called modifying factors, and that the change in the hooded pattern was the result of selection among the alternatives presented by these modifiers, of those types which developed larger or smaller pigmented areas respectively. A crucial experiment was devised to decide between these possibilities. Rats of both selected lines were bred back to unselected selfs, the young were inbred, and the hooded pattern was recovered in the grandchildren; if the modification had taken place in the hooded gene the recovered hooded rats would have received fully modified hooded genes, and must have been as dark, or as light, as the hooded line from which they were obtained; but, if other factors were responsible, the hooded grand-children would have received these equally from their self and from their hooded grand-parents, and would consequently be less extremely dark or light than the latter. The second alternative was proved to be correct, the modification being readily transmitted by self rats which contained no hooded gene. The gene, then, may be taken to be uninfluenced by selection, but its external effect may be influenced, apparently to any extent, by means of the selection of modifying factors.

Unless these analogies are wholly misleading, we should suppose that the factors H and R which Fryer found to determine the differences between the polymorphic forms of *P. polytes*, each arose suddenly by a mutation, and that the new genes so produced have been entirely unmodified since their first appearances. On the other hand, we should see no reason whatever on genetic grounds to believe that the combination HHrr on its first appearance at all closely resembled the modern form *polytes*, or was an effective mimic of *P. aristolochiae*; nor that the combination HHRR resembled the modern form *romulus*, or was an effective mimic of *P. hector*. The gradual evolution of such mimetic resemblances is just what we should expect if the modifying factors, which always seem to be available in abundance, were subjected to the selection of birds or other predators.

Stability of the gene-ratio

It should be emphasized that there is nothing in the argument developed above which helps to explain polymorphism itself. The phenomenon is sufficiently uncommon to suggest that it must always owe its origin to some rather special circumstances; however, the Mendelian character of the phenomenon does suggest one short step in the direction of a solution, namely, that the underlying condition for its development is that the proportionate numbers of the genes of some Mendelian factor, having a fairly marked effect, should be in stable equilibrium, such as that considered in Chapter V.

Stabilizing selection can scarcely be other than exceptional, yet it may be expected to arise in several ways. A Batesian mimic, for example, will receive less protection, the more numerous it is in comparison with its model; a dimorphic Batesian mimic will therefore adjust the numbers of its two forms, if these are dependent upon a single Mendelian factor, until they receive equal protection; any increase in the numbers of one form at the expense of the other would diminish the advantage of the former and increase that of the latter, thus producing a selective action tending to restore the original proportion. A mimic, owing its advantage to Müllerian situations only, should not be dimorphic unless additional causes of stability are at work, for apart from these the selection produces an unstable equilibrium, from which the ratio will continue to depart until one or other type is exterminated.

A second form of stabilizing action is found in reproductive selection. The stable ratio of the sexes is clearly due to this cause, as is that of the thrum-eyed and pin-eyed primroses. It is interesting to note that Fryer, in his breeding experiments with *Papilio polytes* observed numerous cases of sterile unions, which suggested to him the possible existence of 'illegitimate' pairings. One of the simplest possibilities of this type is a merely greater fertility of the heterozygous as compared to the homozygous condition.

It should perhaps be noted that Gerould's work, on the dominant white observed in the female of several species of *Colias*, also reveals some peculiar features suggestive of a stability mechanism governing the yellow-white gene-ratio. Gerould reports that great difficulties were encountered in obtaining the homozygous white types, these difficulties being evidently connected with the occurrence of a closely linked lethal factor. When pure white broods had been obtained, from a strain apparently freed from the lethal, the failure of the males to mate caused the introduction of wild males, and these were found to bring in the lethal factor. The fact that this particular lethal is apparently not rare in nature, although we should expect it to die out somewhat rapidly, suggests that a stabilizing system must be present. The genetic complexities are not fully elucidated, for certain types of mating seem regularly to give an abnormal sex-ratio (3 ♀ : 2 ♂). It is interesting in connexion with the modifying effects of selection, that Gerould notes the occurrence of a fluctuating tinge of yellow on the wing of the genetically white female, and ascribes its variability to secondary factors.

Both the white form in *Colias* and the two mimetic forms of *polytes* are confined to the females, and in both species the mutant form is dominant to the older gene from which it presumably arose. This circumstance, together with the probability in both cases of stability of the gene ratio, suggests that the mutant *form* enjoys a selective advantage, and has for this reason been acquired by the heterozygote, and that the selective disadvantage of the mutant gene is confined to the reproductive factor of genetic lethality or sterility associated with the mutant homozygote. In the case of the mimetic forms of *P. polytes*, their selective advantage had already been inferred by students of mimicry, though whether the white *Colias* can be brought under the same principle appears to be doubtful. If, however, this surmise as to the causes of genetic stability should be proved correct, these and analogous cases should provide an unexampled opportunity of actually assessing the magnitude of a particular selective agency in nature. For reproductive causes of selection, especially if dependent on absolute sterility or lethality, should be capable of experimental demonstration and measurement, and their selective efficacy can be calculated from the frequencies with which the alternative genes, or gene combinations, are found in nature. They should then form a measurable standard to which the efficacy of mimetic selection, against which they must be balanced, can be equated, and even its local variation examined. The importance of such a direct determination need scarcely be emphasized; the hindrances to free reproduction in these groups have appeared hitherto merely as an obstacle impeding the Mendelian analysis of the polymorphic forms; it is much to be hoped that, in view of the application outlined above, their elucidation may, in future studies, be made a principal object of research.

Whatever be the cause to which a factor owes its stability, any species in which a stable factor occurs will be potentially dimorphic, and permanently so unless in changed conditions the stability can be upset. If, in this condition, selection favours different modifications of the two genotypes, it is clear that it may become adaptively dimorphic by the cumulative selection of modifying factors, without alteration of the single-factor mechanism by which the dimorphism is maintained.

Summary

The theory of mimicry is of special interest for the student of Natural Selection, as affording examples of the adaptive significance of specific and varietal differences, and by reason of the great disparity between the views formed by the pioneers of Mendelism and those of selectionists.

The theory of Bates involves errors only of the senses of the predator; it is subject to limitations imposed by the evolutionary reactions of both predator and model. The theory of Müller involves errors both of the senses and of the judgement; the predator and both species of prey are all subject to evolutionary tendencies favourable to Müllerian mimicry.

The statistical limitation of Müllerian mimicry put forward by Marshall is based on insecure reasoning, and on examination is found to be wholly imaginary.

The observational bases for mimicry theory may be analysed into five distinctive categories, two of which are characteristic of Natural Selection generally. Although full evidence is available only for a very small minority of the cases suspected, the concurrence of independent classes of observations puts the well-investigated cases beyond possibility of doubt.

The evolution of distastefulness presents a special problem, which may find its solution in the protection afforded by special distastefulness among gregarious larvae to members of the same brood.

The tendency of Müllerian selection to cause the gradual ·and mutual approach of different warning patterns supplies an explanation of the difficulty felt by Punnett with respect to mimicry rings. His further argument from the Mendelian character of polymorphic mimics that these mimics arose as such by single saltations, overlooks the effects of other factors in modifying the forms determined by the original mutations.

Mendelian polymorphism normally implies a stable gene-ratio in the determining factor. If the mechanism of stability involves, as one may suspect in *Papilio polytes* and *Colias*, an element of sterility or lethality balanced against an element of bionomic advantage, it should be possible by analysing the mechanism of stability to evaluate the intensity of selective advantage experienced in nature by the favoured types.

VIII

MAN AND SOCIETY

On Man, prominence of preliminary studies. The decay of civilizations. Sociological views. Insect communities. Summary.

. . . and if her wretched captives could not solve and interpret these riddles, she with great cruelty fell upon them, in their hesitation and confusion, and tore them to pieces. BACON (*Sphinx or Science*).

On Man

THE earlier discussions of Man in connexion with evolutionary theory were principally devoted to the establishment of two points. First, that man, like the other animals, owed his origin to an evolutionary process governed by natural law; and next that those mental and moral qualities most peculiar to mankind were analogous, in their nature, to the mental and moral qualities of animals; and in their mode of inheritance, to the characters of the human and animal body. However necessarily the second conclusion may seem to follow from the first on any unified view of organic nature, the fact that man's traditional opinion of himself constituted the main difficulty to such a unified view caused the researches, which have led to the final acceptance of Darwin's conclusion on this matter, to follow two paths, distinct in subject and method, which may be typified by the labours of Huxley and of Galton. The similarity of the human to other animal bodies must have been obvious from the earliest times. On the crudest scientific classification, he must be placed as one of a particular order of mammals. An enormous mass of investigation of the form, development, and workings of his body, is found to be consonant in every detail with the view that he is an old world monkey, most closely allied to the tailless apes.

The minuteness with which this subject has been examined would scarcely have been necessary, from the evolutionary standpoint, but for the difficulties presented by the human mind. These difficulties are of two kinds. The first flows from the great development of the human mind compared with the minds of other animals of which we know enough to make any safe comparison. The second is that we have an interior view or consciousness of the human mind, and find in it qualities of great consequence or value to ourselves, which, so long as we remain men, must appear to us immensely superior

189

to anything else in organic nature. Without this second factor, I cannot think that the mere difference in degree of development would have occasioned any popular difficulty in accepting the scientific standpoint. Many animals show great development of particular organs, without leading to any popular misapprehension of their natural affinities. The trunk of an elephant is, regarded *merely as a structural development*, at least as remarkable as the human brain, though it happens to be less important to us.

Unfortunately for the general adoption of such an objective view of things human, as is necessary for the scientific understanding of human affairs, it is often felt to be derogatory to human nature, and especially to such attributes as man most highly values—as if I had said that the human brain was not more important than the trunk of an elephant, or as if I had said that it ought not to be more important to us, if only we were as rational as we should be. These statements would be unnecessarily provocative: in addition they are scientifically void. And lest there should be any doubt upon a matter, which does not in the least concern science, I may add that, being a man myself, I have never had the least doubt as to the importance of the human race, of their mental and moral characteristics, and in particular of human intellect, honour, love, generosity and saintliness, wherever these precious qualities may be recognized. The supreme value which, I feel, ought to be attached to these several aspects of human excellence, appears to provide no good reason for asserting, as is sometimes done, with a petulant indignation not unmixed with spiritual arrogance, that such a low matter as natural causation cannot be of importance to these sublime things. On the contrary, it introduces the strongest motive for striving to know, as accurately and distinctly as possible, in what ways natural causes have acted in their evolutionary upbuilding, and do now act in making them more or less abundant.

This reasonable, and I believe inevitable, consideration has been too frequently neglected by writers on both sides of the controversy. Some seem certainly to have thought that the replacement, by rational, of sentimental or superstitious beliefs, on matters of fact, could be paralleled, and as it seemed completed, by the establishment of moral and aesthetic valuations upon an exclusively scientific basis. Others, whose attitude can be explained, rather than excused, by obscurantist opposition, would seem to have developed a positive dislike for the higher qualities of the human consciousness, at least for controversial purposes, and to delight in assimilating the poetic and religious aspirations of mankind to whatever unedifying proclivities may happen to have a similar physiological basis. It seems to be not uncommon for medical students, who have learnt something of the psychological effects of the sex hormones, to feel that the poetic emotions, which the rest of mankind associate with sexual love, are thereby discredited. And, although this may be dismissed as an extreme and juvenile aberration, men are in general somewhat reluctant to give up the early accepted dogma that whatever is noblest in their nature must be due to causes, and even arise by processes, which in themselves possess the same value. Perhaps the basis of this dogma, which seems to be, in itself, indefensible, may be found in the aristocratic tradition by which men were valued at least as highly for their ancestry as for their personal qualities. Whatever may be its cause, the intellectual efforts made by a man such as Wallace, to avoid ascribing the higher human faculties to natural causation, should warn us of a source of error so powerful that it might perhaps have permanently incapacitated mankind from furthering the development of what he values most.

The conscious and active being is concerned with himself habitually as an original cause. If he considers the natural causes which have brought himself into existence his perspective is inverted, and his intuitions fail him. It is readily to be understood that differences in behaviour, whether due to conscious deliberation or to impulsive reaction, do in fact determine differences in the rates of death and reproduction. Nor can we doubt that these differences in behaviour flow from personal individuality in the constitution of the mind, without which all men would act alike and the concept of voluntary action would be an illusion. It is true that this concept implies that we take circumstances into consideration and act, or refrain from acting, accordingly; but also more emphatically that *we* weigh the circumstances, and that others in like case might have acted differently. Apparent as is this aspect of voluntary action it is a remarkable fact that many, who would find no difficulty in conceiving the involuntary reactions of mankind to be modified by a selective process, yet find a difficulty in applying the same argument to matters involving voluntary choice. I suspect the reason to be that voluntary choice is open to modification, apart from other circumstances, by persuasion, and that the importance of the effects achieved by this means sometimes prevent us from perceiving very striking contrasts in the ease with which persuasion is effected.

The supreme inner arbiter of our choice in matters of right and wrong we call our conscience. We do not regard our consciences as open to outside persuasion, and from this point of view they must be regarded as wholly innate. We do, however, show some ambiguity in the extent to which we recognize their individuality. Emphatically your conscience is no substitute for my conscience; yet, we are sometimes tempted to be so uncharitable as to imply that, apart from hypocrisy or an acquired callousness or blindness towards their own better nature, others would assuredly comply with our own standards. From the standpoint of selection it is a matter of indifference whether we regard mankind as differing in their very consciences, or merely in their reactions to their inward monitor. Since, however, each man is aware only of his own conscience I shall be content, in what follows, to take the former view.

However obvious it may seem to some, it is certainly necessary to insist upon the point that the systematic position of mankind and the demonstrated inheritance of the mental and moral qualities do not exhaust the evolutionary interest of the human species. On the contrary, these are only preliminary inquiries designed to examine whether man, and if so, whether man in all his aspects, falls within the scope of a naturalistic theory of evolution. They are preliminaries to our interest in the evolutionary history and destiny of mankind, a subject the interest of which seems to have been very inadequately explored. An animal which in its comparatively recent history has undergone profound changes in its habitat, diet, and habitual posture, should, for these reasons alone, be of sufficient interest to the evolutionist. If we add the development of a social organization unparalleled within its own class, the use of artificially constructed implements, and a means of expressing and recording its experiences and ideas, it is obvious that to the non-human observer mankind would present a number of highly interesting evolutionary inquiries and would raise questions not easily to be answered only by the use of comparisons and analogies.

Most modern writers on genetics seem, with respect to man, to fall into two opposite, but equally unsatisfactory, attitudes. A minority appear to fear that the purity of their subject, as an abstract science, would be contaminated were it applied to the species to which they themselves belong; and although, perhaps, interested in the practical improvement of domesticated animals and plants, are careful, if any point in human genetics has to be mentioned, to dissociate themselves from any suggestion that it also may have practically important consequences. A larger, and more enterprising school, fully imbued with a sense of the universal applicability of genetic knowledge, recognize in mankind a useful field for its exploitation. They are more ready, and are perhaps better prepared, to appreciate the similarities of human inheritance to that of *Drosophila* or Maize, than they are to appreciate the special problems which the evolution of man in society presents; and will sum up the human problem in a cursory, and even superficial, chapter at the end of an elaborate, and often admirable exposition of modern genetic discoveries. While genetic knowledge is essential for the clarity it introduces into the subject, the causes of the evolutionary changes in progress can only be resolved by an appeal to sociological and even historical facts. These should at least be sufficiently available to reveal the more powerful agencies at work in the modification of mankind.

The decay of civilizations

The decay and fall of civilizations, including not only the historic examples of the Graeco-Roman and Islamic civilizations, but also those of prehistoric times, which have been shown to have preceded them, offers to the sociologist a very special and definite problem—so sharply indeed that its existence appears to challenge any claim we dare make to understand the nature and workings of human society.

To be used properly the term civilization must be applied not merely to those societies the institutions of which we see reason to admire, but also to the aggregate of all the social adaptations appropriate to the permanent existence of a dense population. In general form these adaptations have a universal character. In all societies which we call civilized, the personal understandings, which a man can form with a small circle of immediate acquaintances, are supplemented by a vastly more numerous system of conventional understandings, which establishes his customary relations, his rights and his obligations, with regard to the entire society of which he is a member. He is thus free to devote himself to productive labour even of a highly specialized character, with confidence that the production of his primary necessities, the protection of his possessions from violence, and even the satisfaction of his moral and intellectual needs, will be undertaken by the labours of others, who make of such tasks their special occupations. The specialization of occupations, involving the customary acceptance of a conventional standard of exchange, the maintenance of public order, and the national organization of military preparations, are thus the universal characteristics of the civilized in contradistinction to the uncivilized societies of mankind. It is a matter of experience, which no one thinks of denying, that such an organization does in fact enable a given area to support a much larger population, and that at a higher level of material and intellectual well-being, than the uncivilized peoples who could alternatively occupy the same territory.

Such a civilized society, once organized and established, how is it possible to imagine that it should fail in competition with its uncivilized neighbours ? The latter occupy their territory more sparsely, they lack moreover the organized central government which could mobilize to advantage their scanty numbers. On the contrary, our experience of uncivilized populations shows them to be divided by hereditary enmities and petty jealousies, which should make their union, even upon a question of the simplest national interest, almost an impossibility. Industrial organization gives to civilized peoples, in antiquity as well as in our own times, the advantage of superior weapons; while the habits of co-operative labour enable them to adopt a more regular, co-ordinated, and effective system of military tactics. Above all, the superior knowledge which a civilized people can, and does indeed, continually accumulate, should enable them to act generally with superior information, with a surer foresight of the consequences of their collective action, and with the capacity to profit by experience, and to improve their methods if their first attempts should prove to be unfortunate. Bearing in mind the unquestionable advantages of superior knowledge, of co-ordinated efforts and of industrial skill, should we not confidently anticipate, if we were ignorant of the actual history of our planet, that the history of civilization would consist of an unbroken series of triumphs ; and that once the germ of an organized society had made its appearance, in Babylonia, perhaps, or in Egypt, it would be only a question of time for every country in the world to be in turn absorbed and organized by the Babylonian, or Egyptian, civilization ?

The indications which we possess of the earlier civilizations, as well as the plain narrative of the historic period, differ remarkably from the rational anticipation deduced above. We see, indeed, a certain tenuous continuity in many elements of traditional civilization which are handed on from one social group to another, as these in turn become civilized. But this circumstance seldom even obscures the contrasts between the social groups, involving differences between the territory, language, religion, and race, associated with the highest civilization at different epochs. These contrasts are obviously associated, in the case of each great change, with the violent irruption of some new people of uncivilized origin. The experiment of becoming civilized has, in fact, been performed repeatedly, by peoples of very different races, nearly always, perhaps, with some aid from the traditional ideas of peoples previously civilized, but developing their national and industrial organizations by their own progressive powers; and in all cases without exception, if we set aside the incomplete experiment of our own civilization, after a period of glory and domination accompanied by notable contributions to the sciences and the arts, they have failed, not only to maintain their national superiority, but even to establish a permanent mediocrity among the nations of the globe, and in many cases have left no other record of their existence than that which we owe to the labours of archaeologists.

Before considering those causes which I propose to assign both to the phenomenon in general, and to its more salient characteristics, in the course of the succeeding chapters on Man, it is as well to give some attention to the preliminary question: Of what sort should be an explanation which we should regard as adequate? It is easy to find among the peoples of antiquity institutions disagreeable enough to our modern feelings, it is easy to criticize their educational ideas, or the forms of government which they have successively adopted; above all it is easy to find fault with their ignorance of economic law, and to ascribe to their mistakes in this domain the same civil and political misfortunes, which we anticipate equally from the parallel errors of our political opponents! Such arguments are not only inconclusive from our ignorance of the laws of causation upon which they rely, they are also demonstrably insufficient to meet the requirements of our special problem. For, in the first place, our knowledge of the earlier stages of the history of great peoples shows us customs no less repulsive, manners no less licentious, a neglect of education at least equally pronounced, and ignorance of economic law as absolute as any which can be ascribed to their civilized successors. In the second place, moreover, in a period of national decay it would be unreasonable to expect that any aspect of national life, political, religious, intellectual, or economic, should remain in a healthy and flourishing condition, or that the misfortunes of the times will escape the complaints of observant contemporaries. That the condition of agriculture, for example, was unsatisfactory in the later Roman empire, though a legitimate inference from the state of that society, fails to constitute in any useful sense an explanation of its progressive decay. Peoples in the prime of their powers appear to find no difficulty in making good use of very inferior natural resources, and adapt their national organization with complete success to much more violent changes than those which can be adduced to explain the misfortunes of the later stages of their civilization.

A physician observing a number of patients to sicken and die in similar though not identical conditions, and with similar though not identical symptoms, would surely make an initial error if he did not seek for a single common cause of the disorder. The complexity of the symptoms, and of the disturbances of the various organs of the body, should not lead him to assume that the original cause, or the appropriate remedial measures, must be equally complex. Is this not because the physician assumes that the workings of the body, though immensely complex, are self-regulatory and capable of a normal corrective response to all ordinary disturbances; while only a small number of disturbances of an exceptional kind meet with no effective response and cause severe illness ? It is impossible to doubt that we have equally a right to expect self-regulatory power in human societies. If not, we should be led to predict that such societies should break down under the influence of any of the innumerable accidents to which they are exposed. Uncivilized societies of various kinds have adapted themselves to every climate, from the Arctic, to the forests and deserts of the Tropics. They share the territories of the most savage or poisonous animals, and often long withstand without disruption the assaults of most implacable human enemies. Social progress has not been arrested by the introduction of new weapons, of alcohol, or of opium, or even of infanticide ; yet these introductions might each of them seem to threaten the existence of the race. That civilized men, possessed of more effective appliances, with access to more knowledge, and organized for the most detailed co-operation, should prove themselves incapable of effective response to any disturbance of their social organization, surely demands some very special explanation.

Sociological views

A philosophical view of history, which has attained to great popularity in several continental countries, regards the rise as well as the fall of civilizations as but the successive phases of a cycle of growth and decay, which, it is supposed, repeats itself, and will repeat itself endlessly, throughout human history. So long as we aim at no more than an effective generalized description of the phenomena there is much to be said for considering the phases of rise and development of the great civilizations in conjunction with the phases of their decay and collapse. It is a real advantage that the parallelisms between the earlier phases of different cultures should be recognized, and that their relations in time to recognizably parallel later phases should be established as accurately as possible. Generalized description should, however, never be regarded as an aim in itself. It is at best a means towards apprehending the causal processes which have given rise to the phenomena observed. Beyond a certain point it can only be pursued at the cost of omitting or ignoring real discrepancies of detail, which, if the causes were understood, might be details of great consequence. Alternatively, somewhat different states and events are subsumed under generalized and abstract terms, which, the more they are made comprehensive, tend to possess the less real and definable content. Finally, any purely descriptive general picture of events in time is in its nature fatalistic and allows no place for intelligent and corrective intervention.

The early evolutionary speculations of the Greeks had this fatalistic and sterile character, for the lack of any clearly understood mode of causation which could bring about the modification of living organisms. It is characteristic of the scientific attitude of the middle of the nineteenth century that the evolutionary theories of earlier generations excited little interest in the absence of a satisfactory explanation capable of expressing the means of modification in terms of known causes. Even Darwin and Wallace, although Darwin at least was in possession of much indirect evidence, did not put forward their evolutionary theories until each had satisfied himself that, in the struggle for existence, natural selection did provide the efficient cause. With a clear grasp of scientific principle, which is not always sufficiently appreciated, it is evident that they felt that the mere historical fact of descent with modification, however great its popular interest, could not usefully be discussed prior to the establishment of the means by which such modification may be brought about. Once the nexus of detailed causation was established, evolution became not merely History, but Science. Evidently, the more special and peculiar is the case considered, the less can descriptive analogy be relied upon, and the more essential is a knowledge of the laws of causation.

In the descriptive study of the rise and decay of human civilizations the closest analogy which is found with any process capable of scientific study is that with the growth and death of the individual organism. In both, it may be said, we find, in regular sequence, a period of developmental vigour and superfluous vitality with the acquisition of new powers, a period of maturity with substantial achievement, a period of stagnation in which the organism or the society can do little more than hold its own, and a period of decay followed sooner or later by dissolution. The moment we attempt, however, to interpret this generalized description in terms of known causes, the analogy is seen to be as false as it well could be. The 'youthful' peoples whom we see at the dawn of their civilized history have already behind them a social history far longer than the civilized period before them. They have not recently developed from elements secreted by previous civilizations, elements which might be supposed to carry the hereditary determiners of the cycle of changes to be enacted. The healthy society may be said to grow and to assimilate; it cannot be said to reproduce itself. And without reproduction there is no *terminus a quo* for the sequence of growth and death. Societies are potentially immortal, bearing within themselves the power continually to replace every living element in their structure. The phenomenon of senescence from which the whole analogy arose is only observed in societies which have for some little time enjoyed that closely co-operative structure which we call civilization.

If we set aside the purely descriptive point of view, it is apparent that, in associating the rise and growth of civilizations in a single sequence with their decay and fall, the nature of the problem has only been obscured. Obviously the phenomena of decadence can only be presented after a certain level of success has already been achieved. But, whereas the decay of civilizations presents an abrupt and un-expected problem, the advantages to be reaped from the progressive adoption of those phases of culture which we call civilized are, to all, obvious and familiar. Consequently, while a solution of the problem presented by decadence in terms of the detailed sociological reactions to which it is due, would doubtless throw light on sociological causa-tion in general, and consequently on the characteristics of the earlier as well as of the later phases of civilized society, the mere fact that a society, as it acquires the arts and organization of civilized life, simultaneously experiences a great accession of prosperity and power, is itself due to causes which are perfectly familiar and fully under-stood. These advantages would be even further enhanced and pro-longed, if, as it was formerly thought could be safely assumed, life in the civilized condition, as in the barbaric state, favoured the sur-vival and reproduction of those human types who could most effectively promote the prosperity of their society, and who on the other hand were most apt temperamentally to appreciate and exploit its advantages. The evolution of man from savagery, through bar-barism to a final highly civilized condition might then be regarded, as by Herbert Spencer, as forming a continuous and inevitable process of human advancement. In the light of this highly rational optimism the failure of high civilizations in the past appears as a single and formidable problem.

Insect communities

The only animal societies in which co-operation is sufficiently highly developed to justify comparison with civilized men are those of the social insects. In these, specialization for the performance of various tasks such as defence, building, direct social services for the benefit of the rising generation, and the collection or cultivation of food, is sufficiently clearly developed; nevertheless, although social life has appeared independently in several distinct groups, all alike present the same initial difficulty to comparison with human societies, in that, in addition to showing such specialization in behaviour or capacity as is necessary for organized co-operation, all display an apparently superfluous specialization in reproduction. A single queen termite, to take an extreme example, is said to maintain the wastage among a whole society of many millions of workers, by producing continuously an egg every few seconds. In general, the vast majority of a community of social insects take no part in reproduction; and while it cannot be denied that some small economic advantage accompanies this, like other specializations for the division of labour, yet the danger to the community of entrusting its future existence to a single life is obvious and serious. In this respect the insect society more resembles a single animal body than a human society, for although many tissues are capable by cell division of the replacement of damaged parts, yet the reproduction of the whole organism is confined to specialized reproductive tissue, whilst the remainder of the body with its various co-operative functions, co-operating with but taking no part in reproduction, is in this respect analogous to the body of sterile workers which constitutes the bulk of the hive. It is of some interest to follow out the effects of the system of reproduction adopted by the social insects in contrast to the individualistic system of reproduction in human societies.

Human communities show the same kind of genetic variation as is shown by the species as a whole; individuals are fitted for their various special tasks, partly by direct and deliberate selection, partly by indirect selection *via* social class, and partly by special education, including in that term not only paid instruction in youth, but also experience gained in practice. The different tissues of a metazoan body, are, it seems, normally identical genetically, and owe the differences which they exhibit entirely to differences in the influences to which they are exposed from other developing tissues. The different offspring of a single queen cannot be genetically identical, for the factors in which their mother is heterozygous must segregate, so that in each such factor offspring of the two kinds are produced in equal numbers. Such genetic variability, however, must be different in each different society, and seems not to be utilized in producing polymorphism among the workers, or even between the sterile and the reproductive types. Although the subject is very obscure, the predominant weight of opinion appears to favour the view of Wheeler that the principle controlling influence is exercised through the quantity of food supplied to the larvae during development. The mechanism by which polymorphism is produced is thus very different from the simple Mendelian mechanism found in the polymorphic butterflies, and one far more appropriate to the needs of social insects. At the same time it should be noticed, as with polymorphic butterflies, that the particular mechanism employed to secure poly-morphism throws no light whatever upon the adaptive and evolutionary significance of the polymorphism itself. The manner in which developmental changes are modified by restricted nutrition is a variable equally exposed to selective action as the manner in which a wing pattern is modified by a particular Mendelian mutant. The selection in this case must act exclusively upon the reproductive insects *via* the prosperity of the society from which they arise; and although the effect of such selection may be to modify only the sterile workers, this presents no more difficulty than that a selection acting exclusively upon the gametes of a sheep, *via* the observable characteristics of the animal which bears them, should modify the nature of its wool.

Although real genetic differences must exist among the sterile workers of the same society, these differences are without selective effects. The relative frequencies with which the different genotypes appear among the offspring of a single queen must be presumed to remain constant during her reproductive period, and the selection of males and females for reproduction depends only on their individual qualities, and the aggregate quality of the communities from which they arise. On the contrary, the genotypic differences which characterize the individuals of societies practising individualistic reproduction are exposed through their differential rates of death and reproduction to an intra-communal selection capable of modifying indefinitely the genotypic composition of the body politic. Among a group of small independent competing tribes the elimination of tribes containing an undue proportion of the socially incompetent, and their replacement by branches of the more successful tribes, may serve materially to maintain the average standard of competence appropriate to that state of society. Even in such a state of society intra-communal selection will undoubtedly be at work, and the larger the social group becomes the greater is the importance of the intra-communal element. Consequently, from the earliest times of which we have knowledge, the hereditary proclivities, which undoubtedly form the basis of man's fitness for social life, are found to be supplemented by an economic system, which, diverse as are the opinions which different writers have formed about it, appears to the writer to be one of the unconscious triumphs of early human organization.

The basis of the economic system consists in the free interchange of goods or services between different individuals whenever such interchange appears to both parties to be advantageous. It is essential to the freedom of such agreements that the arbitrary coercion of one individual by another shall be prohibited, while, on the other hand, the coercive enforcement of obligations freely undertaken shall be supported by the public power. It is equally essential that the private possession of property, representing, as in this system it must do, the accumulation of services already performed to other members of the society, and the effective means of calling upon equivalent return services in the future, shall be rigorously protected. In the theory of this system each individual is induced, by enlightened self-interest, to exert himself actively in whatever ways may be serviceable to others, and to discover by his ingenuity new ways or improved methods of making himself valuable to the commonwealth. Such individuals as succeed best in performing valuable services will receive the highest rewards, including, in an important degree, the power to direct the services of others in whatever ways seem to them most advantageous. Those, on the contrary, who fail most completely to perform socially advantageous actions have the least claim upon the wealth and amenities of the community. In theory they may perish of starvation, or may become indebted up to the amount of the entire potential services of the remainder of their lives, or of the lives of their children.

It need scarcely be said that this economic system has never formed the exclusive basis of social co-operation in man. It has at most been partially established in compromise with customary social sentiments already in being, founded in a time of less closely co-operative societies. Nevertheless, it bears a sufficient resemblance, both to the theory of rationalistic economists, and to the practice of various ancient civilizations, to indicate that we have presented, in an abstract and ideal form, a real and effective factor in human social organization. The biological importance of this factor lies in the safeguard which it appears to provide that intra-communal selection in human societies shall not favour the multiplication of unproductive or parasitic types, at the expense of those who exert themselves successfully for the common good. On the contrary, it seems to insure that those who produce the best goods or provide the most valuable services shall be continually augmented in each succeeding generation, while those who, by capacity or disposition are unable to produce

goods equivalent to what they consume, shall be continually eliminated. Nor is this beneficial selection confined to individuals in their
capacity of producers. In consumption and distribution an equally
beneficial selection would seem to be in progress. The individual who,
by reason of his imperfect instincts, is tempted to expend his resources in ways which are not to his biological advantage, the individual who from prejudice favours a bad market, or who is temperamentally incompetent in striking a bargain, is equally at an economic
and, it would seem, at a selective, disadvantage. This selection of
the consumer provides in an important respect the theoretical com-
pletion of the individualistic economic system, for it supplies a means
by which the opportunities of gaining wealth by the provision of
illusory benefits, shall become ever narrower, until all substantial
sources of profit are confined to the provision of real public benefits. The
population produced by such a system should become ingenious and
energetic industrialists, shrewd and keen in the assessment of social
value, and with standards of well-being perfectly attuned to their
biological and reproductive interests. To complete the picture, at
the expense of anticipating a little a subsequent argument, our
economic Utopians must be endowed with consciences which recognize the possession of wealth, at least as a means to reproduction, as
the highest good, and its pursuit as the synthesis of all virtuous
endeavour. To them the wealthy man would enjoy not only the
rewards, but also the proofs of his own virtue, and that of his forbears ;
he would be in some sort a saint, to co-operate in whose virtuous
proceedings would be a supreme felicity. Upon such men, no public
honour could be bestowed more noble than a direct cash payment,
and to purchase other honours for money would seem not so much
corrupt as insane. Charity, in the sense of the uneconomic relief of
poverty, would evidently be a vicious weakness, although there would
be some virtue in shrewdly backing for mutual advantage the capable,
but accidentally unfortunate.

If in practice man's sociological development had indeed taken
this turn it might have been said that man, by individualizing
property, had found the appropriate social concomitant for his individualistic habit in reproduction ; that he had established a type of
society unattainable by any order of insects, for lack of an intellectual
equipment adequate to deal, with the necessary justice and con-
sistency, with disputed ownership or contractual obligations—for

lack, in fact, of judges and lawyers; and that for this reason the only means open to them, to avoid the disastrous consequences of intra-communal selection, lay in the complete elimination of competition between reproductive members of the same society, by the concentration of the whole duty of reproduction upon a single individual. As the matter stands, however, it appears that even if an insect community had achieved some efficacious system of instincts, which would produce the economic effects of personal ownership, its difficulties would not necessarily have been at an end; for the instinctive feelings and prejudices of social man do not seem at all to have developed in the direction of a more strictly economic and less 'sentimental' basis for social institutions. Whether an imaginary insect could, and whether man himself can, overcome the further sociological difficulty which has arisen can only be decided when that difficulty has been more thoroughly examined. The important point for the present chapter is that the difficulties that man has encountered and, at most, partially overcome, have been evaded in all orders of social insects, by forming the community from the offspring of a single individual.

The primary evolutionary process for us to notice here is the enormous, though doubtless gradual, increase in individual fertility in insects. As was remarked in Chapter II, the importance of Natural Selection in moulding fertility to its optimum value has been obscured by the importance attached, by early evolutionary writers, to a 'high rate of increase' as the primary driving force of Natural Selection itself. When, on the contrary, we regard fertility as a secondary phenomenon, determined by Natural Selection, it becomes important to consider what circumstances influence the level of fertility which is, in any particular case, optimal. In this connexion Major Leonard Darwin has pointed out that the principal importance should be given to the factor of parental care, including in that term all expenditure in the form of nutriment, effort, or exposure to danger, incurred in the production and nurture of the young. In organisms in which that degree of parental expenditure, which yields the highest proportionate probability of survival, is large compared to the resources available, the optimal fertility will be relatively low. Any circumstance which materially lightens the burden on the parent will necessarily have an immediate effect in favouring survival; it is more important that it will usually lower the optimal expenditure, and consequently tend to raise the level of fertility. Major

Darwin illustrates this principle by the example of the parasitic cuckoos, which have found a way to relieve themselves of a great part of the parental burden borne by other birds, and have, in fact, acquired a considerably greater fecundity than their non-parasitic allies. The social insects offer an even more striking example of the same principle. As soon as the young adults of any incipient social form took either to performing the preparatory labour for reproduction, or to tending the young, before they had themselves commenced to reproduce, the balance of selective advantage would have been shifted towards favouring the fertility of the foundress of the colony, and favouring equally the development of the organs and instincts of workers rather than of queens among her earlier, and possibly less well nourished, offspring. The enormous development of fertility alongside extensive sterility in all groups of social insects shows how powerful is the action of selection in modifying this particular character.

In the three following chapters we shall examine, first, the objective evidence as to the magnitude and heritability of variation in human fertility; next, we shall consider some of the widespread evidence of its association with social class; and, finally, we shall put forward a theory of the selective process by which this association appears to have been established. On this theory it may be seen that its destructive consequences are not incapable of rational control.

Summary

The earlier evolutionary work on Man was naturally directed towards establishing the two preliminary points, (i) that Man like other animals has arisen by an evolutionary process, (ii) that the mental and moral qualities of Man are equally with the physical qualities the result of natural causation. The difficulty felt by Man in readjusting his traditional opinion about himself has required that these two points should have been established with considerable cogency.

A naturalistic view of Man provides no means of putting on an objective basis those mental valuations, moral and aesthetic, to which Man attaches such high importance; it cannot, therefore, be used to throw doubt upon these valuations. The high human value of the moral attributes does, however, invest their natural causation with a special importance.

Among the problems presented by the social evolution of Man the most conspicuous is that of the decay and ruin of all civilizations previous to our own, in spite of their having had every reason to anticipate continued success and advancement. The purely descriptive treatment of the rise and fall of civilizations is inadequate without an examination of the operative causes to which changes in social structure are due.

In the animal kingdom the only societies which bear comparison with human civilization are those of the social insects, and these all differ from human societies in the striking feature of reproductive specialization. This has the effect of eliminating intra-communal selection as an evolutionary process, so that the dangers to which society is exposed from this source are special to Man. Human societies from very early times have adopted an economic system of individualizing property which might have been expected to control intra-communal selection along socially advantageous lines. The logical effects to be anticipated have, however, not been realized.

The specialization of reproduction in insect communities has been possible owing to the efficacy of selection in modifying fertility. The inheritance of fertility will be seen to be equally influential, though in a very different way, in the evolution of human societies.

IX

THE INHERITANCE OF HUMAN FERTILITY

The great variability of human reproduction. The mental and moral qualities determining reproduction. Direct evidence of the inheritance of fertility. The evolution of the conscience respecting voluntary reproduction. Analogies of animal instinct and immunity to disease. Summary.

> *Or who is he so fond will be the tomb*
> *Of his self-love, to stop posterity?*   SHAKESPEARE.

The great variability of human reproduction

THE existence of statistical data for civilized man might suggest that these should provide an opportunity for evaluating, in this species, the probabilities of any individual leaving 0, 1, 2, . . . offspring, according to the chances of his life as considered in Chapter IV. The case of man will be shown to differ from that of most other organisms, in the high relative importance of differences in fertility and in the comparative unimportance of the differences of mortality between birth and the reproductive period. In view of these characteristics of mankind it would be of very special interest to be able to evaluate, even approximately, the magnitude of the effects of the chance element in causing variation in the numbers of children actually born. Two special obstacles must be encountered in such an attempt, which in the present state of our knowledge can be at best only partially overcome. In the first place the actual variation in the number of children born, let us say, to a group of women at the end of their reproductive period, cannot, with the faintest show of reason, be ascribed wholly to chance. In this respect probably mankind, in the civilized condition, differs materially from all other organisms except the social insects. For unless the rates of reproductive selection in these other organisms are materially higher than we have had reason to suppose, the chance factor must, as was seen in Chapter IV, be the dominant one in individual survival, contributing in fact nearly the whole of the individual variance. In man, on the other hand, we must be prepared to ascribe to differences in individual temperament and disposition an amount of the variance almost as great as that ascribable to chance, and therefore of an altogether different order of magnitude from the reproductive selections to be anticipated in other organisms.

If we cannot evaluate the chance component of the variance by equating it to the total, the alternative method of evaluating it *a priori*, as due to the successive occurrence of independent events, is rendered, at best, extremely uncertain by the long duration and persistence of human purpose. The occurrences of successive children are not independent events in so far as they are conditioned by the married or unmarried state of their parents, although that state may be held to be influenced, though in a very minor degree, by a genuinely fortuitous element. A complete statistical statement of the birth-rates of the married and unmarried at each age, of the frequency with which the unmarried become married, and the married become unmarried by widowhood or divorce, would still afford no direct measure of the extent to which different members of an un-differentiated group, would differ in their number of children. The process of calculation which takes not only the successive births but also marriages, as conditioning the probability of births, as a series of events occurring independently with the appropriate frequencies, seems to err as much in ascribing marriage wholly to chance, as does the simpler method of calculation in which the effect of marriage is ignored.

A. O. Powys (*Biometrika*, IV. 238) gives the number of children born to 10,276 married women dying aged 46 or over in New South Wales from 1898–1902. The number of children varies from 0 to 30, and the number of women recorded at each value are given in Table 7. The average number of children was 6·19, and if the data had represented women of equal natural fertility exposed throughout their lives to equal risk of maternity the variance would certainly be somewhat less than the average number. Actually the variance is 16·18, or considerably more than double the greatest amount ascribable to pure chance.

TABLE 7.

No. of children.	Frequency.	No. of children.	Frequency.	No. of children.	Frequency.
0	1,110	11	568	22	1
1	533	12	422	23	—
2	581	13	226	24	1
3	644	14	129	25	—
4	702	15	57	26	—
5	813	16	39	27	—
6	855	17	12	28	—
7	976	18	5	29	—
8	963	19	2	30	1
9	847	20	2		
10	786	21	1		

Frequency total 10,276.

A great part of the variance must certainly be ascribed to age at marriage. Yet the variance still exceeds the mean very strikingly if data are chosen which should minimize this factor. A group of 2,322 wives married before 20 years of age, and living with their husbands at the time of the Census (1901) 25–30 years later, are distributed as in Table 8.

TABLE 8.

No. of children.	Frequency.	No. of children.	Frequency.	No. of children.	Frequency.
0	44	7	167	14	77
1	41	8	237	15	46
2	63	9	272	16	19
3	65	10	333	17	9
4	83	11	271	18	6
5	105	12	208	19	3
6	141	13	131	20	1

Frequency total 2322

The mean number of children for this group of women, for whom the chosen conditions are especially favourable to fertility, is 8·83, but the variance is still as high as 12·43.

From these examples, and indeed from any series of similar data, it is evident, that while an exact evaluation of the chance factor in reproduction is not possible, yet there clearly exist, both in the age at marriage, and in fertility during marriage, causes of variation so great as to be comparable with the chance factor itself.

The extraordinary variation in fertility in Man has been noticed in a somewhat different manner by Dr. D. Heron, using material provided by the deaths (30,285 males and 21,892 females) recorded in the Commonwealth of Australia for 1912. Heron finds that half of the total number of children come from families of 8 or more, which are supplied by only one-ninth of the men or one-seventh of the women of the previous generation. It would be an overstatement to suggest that the whole of this differential reproduction is selective; a substantial portion of it is certainly due to chance, but on no theory does it seem possible to deny that an equally substantial portion is due to a genuine differential fertility, natural or artificial, among the various types which compose the human population.

The mental and moral qualities determining reproduction

Whereas the part played by chance in producing variations in the rate of reproduction is elusive and difficult to evaluate, that part, which is due to constitutional and therefore hereditary differences in temperament, is manifest to all. When we consider the causes which normally lead to the production of children, the occasions upon which individual temperament is liable to exert a decisive influence will be seen to be very numerous. A considerable percentage of persons of both sexes never marry: the age of marriage is very variable, and with women especially the effect of age is very great: according to the Australian figures of 1911 the maximum rate of reproduction for married women occurs at 18, and at 31 has fallen to half the maximum value. A bride of 30 may expect but 38 per cent. of the family she would have borne had she married at 20, and by 35 the number is further reduced by one-half, and is a little less than 19 per cent. With men the potentiality of fatherhood is usually retained to a considerable age, nevertheless the age at marriage is still very influential, since the most frequent age for the brides increases steadily with the age of the bridegroom. For bridegrooms of from 34 to 44 years of age, brides of the most frequent age are very regularly ten years junior to their mates. Using the age of their wives as a basis for calculation, men marrying at 40 to 44 may be expected to have only two-fifths of the number of children of men marrying 20 years earlier.

The choice between celibacy and marriage, and if marriage be decided upon, between the precipitation and the postponement of the union, is in modern societies very much a matter of temperament. Some men are little charmed by female society, others of a more sociable disposition seem for long unsuited to a permanent alliance: in all normal men the primary impulses of sex appear to be sufficiently developed, but in civilized circumstances the strength of this impulse is not directly exerted towards the conjugal condition: in many men it is overcome by prudence and self denial, in others it is dissipated in unproductive channels. The various privileges and obligations of marriage appeal very differently to different natures: patience, caution, and a strong sense of financial responsibility often postpone, or prevent the initiation of desirable matches: confident, passionate, and impulsive natures marry in circumstances which would awake in the prudent the gravest misgivings. The influence of the innate disposition in this matter is increased in our own times by the absence of any strong social opinion which might regulate the erring fancy of the individual, and by the conditions of urban life

which set no limit to the variety of acquaintance. Upon consideration it would seem that while to the individual, fortuitous circumstances may seem to be of much importance, their influence is easily
exaggerated even in determining the individual destiny. The man
who from an early disappointment condemns himself to celibacy,
is less the victim of misfortune than of his own temperament, as is
the unwary youth who becomes the lifelong victim of an unfortunate
entanglement. In each case temperament largely determines how the
chances of life are taken, and in the aggregate of sufficient numbers
the chance element is wholly negligible.

The initiative in respect of marriage is conventionally regarded as
confined to men, while the factors which determine celibacy or age
at marriage in the case of women are conditioned by the character
and frequency of the offers they receive. Even if this distinction
were absolute in practice, it would be difficult not to admit that
temperamental differences in young women do in fact exert the
predominant influence upon their respective probabilities of marriage
within, say, the next five years. To this probability other hereditary
characteristics will also contribute, in so far as they control beauty,
health or traits of character favoured by suitors. In point of fact,
however, the distinction between the sexes in this matter is far from
absolute and the probability of marriage in either sex is much influenced by the characters, which, on a conventional basis, have been
ascribed to the other.

Besides these causes of varying fertility, which act through the
incidence of the social state of marriage or celibacy, there exist in
civilized and savage life, and have existed from the most ancient
times, a number of practices by which the increase of population is
artificially restricted. Infanticide, feticide or abortion, and the prevention of conception, have all been or are now being practised among
every considerable body of people: to these should be added such
practices as prostitution the use of which attains the same purpose by
competing with instead of corrupting the conjugal condition.

The extent to which infanticide is and has been practised may be
appreciated by a perusal of the masterly section which Westermarck
devotes to the subject. Among uncivilized peoples, though not
universal, it is commonly customary, and frequently compulsory.
Though forbidden by Buddhism and Taoism it is frequent in China.
Among the pre-Islamic Arabs it seems to have been usual and

approved: it is repeatedly condemned in the Koran. It is not mentioned in the Hebrew Scriptures, and may have been unknown to the early Jews. In India where it was forbidden by the laws of Manu, the Rajputs are said to have been particularly addicted to the practice, for which, however, priestly absolution was obtained. In Ancient Greece the practice of exposing healthy infants, though hardly approved, was tolerated, except at Thebes where it was a capital crime. In Rome, though the destruction of deformed infants was enjoined, the murder, and the less decisive act of exposure, of healthy infants was contrary to the prevailing moral standards.

Of those factors affecting the fertility of married persons in our own civilization, by far the chief importance must be given to the prevention of conception. At the present time there can be no doubt of the wide prevalence of birth limitation among married couples in Europe and America. It is earnestly urged in certain restricted circles, that only by the adoption of these methods can private indigence and public scarcity be prevented, and their use is put forward not only as a blessing but as a duty. Public opinion, however, so far as one can gauge it upon a topic, which, however prominently discussed is essentially private, appears to resemble that of Pliny respecting feticide, that it is a venial offence and one that is frequently excusable. ' The great fertility of some women ', says Pliny, ' may require such a licence '. It is important to observe that deliberate contraceptive practices are confidently supposed to be the most important causes, not only of the general decline of the birth-rate in recent decades, but especially of the diminished fertility of the upper classes.

In the practice of, or abstention from, contraception, temperamental differences, including the attitude towards sexual morality generally, are particularly pronounced. This is evident, if only from the various grounds on which this practice is advocated. It is certain that, to some, the freedom from individual liability to the natural consequences of sexual intercourse is a most attractive feature, and that a radical alteration of the accepted attitude towards sexual morality, including the abandonment of the conventional abstinence of the unmarried, would be welcomed. To others, again, who are totally averse from sexual anarchy, a genuine appeal is made through the intensity of the parental instincts; for the prospect of endowing or educating the earlier children may be reasonably held to be impaired by later arrivals. The sentiment that the beneficent effects of parental companionship also gains strength by being concentrated on a few, should perhaps be classed among the excuses, rather than the reasons, for family limitation. Yet even the excuses are instructive from the enormous variability of the emotional response which they evoke. The majority I suspect, even of those that practise it, regard contraception with some degree of reluctance or aversion, but according to the strength of this feeling may be induced to overcome it by a greater or less intensity of economic pressure. Others, again, would certainly feel themselves disgraced if they were to allow economic motives to curtail in any degree the natural fruit of their marriage.

If we consider the immense variety of the temperamental reactions of mankind toward the two predominant factors determining fertility, on the one hand marriage, and on the other the restricted or unrestricted production of children by married couples, it will not be surprising that, as has been seen, an exceptionally great variability ascribable to these two causes should appear in the statistics of civilized peoples; and that the genotypic differences in reproduction outweigh any genotypic differences which it would be reasonable to anticipate in mortality, the more especially since the incidence of mortality among civilized peoples has been much diminished by the advances which have been made in public and private hygiene. In respect to mortality it is necessary to note that only deaths prior to, and in less degree during, the reproductive period, have any selective influence, and that these are already a minority of all deaths. The diagram in Chapter II showing the reproductive value of women according to age gives a fair idea of the importance in this respect of the death-rate at different ages. Even the highest death-rate in this period, that in the first year of life, must be quite unimportant compared with slight differences in reproduction; for the infantile death-rate has been reduced in our country to about seven per cent. of the births, and even a doubling of this rate would make only about a third as much difference to survival as an increase in the family from three children to four.

Direct evidence of the inheritance of fertility

Great as the contribution of fortuitous circumstances to the observed variance in individual reproduction undoubtedly is, it is none the less possible to obtain direct evidence of the importance of hereditary

causes, by comparing the number of children born to individuals
with the number born to their parents. As a biometrical variate the
number of children born to an individual is peculiar in several respects.
It is, like the meristic variates, discontinuous and confined to the
positive whole numbers, while unlike most of these, the frequencies
begin abruptly at zero. Moreover, where reproduction is governed by
the civil institution of marriage, the variate will usually be the same
for a man and for his wife, and cannot be properly regarded as
distinctive of either individually. It is obvious too that where a considerable fraction of the variance is contributed by chance causes,
the variance of any group of individuals will be inflated in comparison with the covariances between related groups, with the result
that all correlations observable will be much diluted in comparison
with those to be found between metrical characters. Rather numerous
groups must therefore be studied if clear results are to be obtained.

From the data given by Pearson of the numbers of children born
to mothers and daughters, in a thousand cases extracted from the
British Peerage, the average number of children born to the daughter
is found to increase according to the size of the family to which she
belongs, from 2·97 from families of 1, to 6·44 for families of 12 or more,
in accordance with the following table:

TABLE 9.

Number of children born to mother.	Number of cases.	Total number of children born to daughters.	Average children per daughter.
1	35	104	2·97
2	67	237	3·54
3	111	249	3·14
4	136	464	3·41
5	138	550	3·99
6	132	513	3·89
7	114	448	3·93
8	87	354	4·07
9	74	313	4·23
10	50	219	4·38
11	24	122	5·08
12	19 ⎫	99 ⎫	5·21 ⎫
13	9 ⎬ 32	77 ⎬ 206	8·56 ⎬ 6·44
14	1 ⎪	10 ⎪	10·00 ⎪
15	3 ⎭	20 ⎭	6·67 ⎭

It will be seen from the averages, which are charted in Fig. 10, that
apart from the inevitable irregularities due to limited numbers, there

is a consistent increase in the number of children born to the daughters, as the size of the family is increased, from which they were derived. On the average an increase of one in the mother's family is followed by an increase of 0·21 in that of the daughter, and although this is

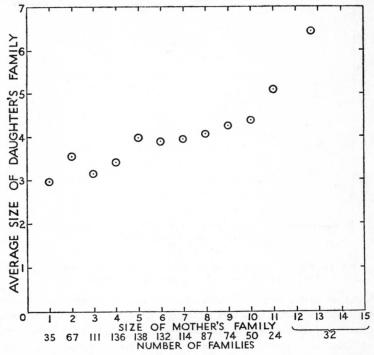

Fig. 10. Average number of children born to peeresses, according to the size of the families of their mother.

less than one-half of what we should obtain in comparing the metrical variates of individual parents and children, when these variates are almost exclusively determined by hereditary causes, yet the value which emerges from the data is more than six times its standard error, and is unquestionably significant of a tendency for the size of a daughter's family to resemble that of her mother in its deviation, whether positive or negative, from the current average of her generation.

It would be important to ascertain to what extent, if at all, this resemblance could be ascribed to a traditional continuity, whether in religious doctrine, moral environment, or social ideas. We could only adopt this explanation, to the exclusion of organic inheritance, by assuming that individuals are not influenced in assimilating such

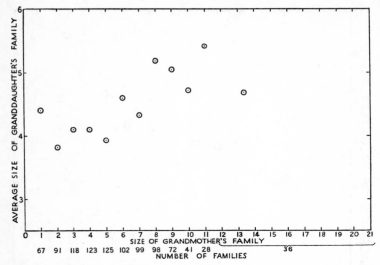

Fig. 11. Average number of children born to granddaughters, according to the number born to their paternal grandmothers.

ideas, or in putting them into practice, in any degree by heritable traits of character, and this is disproved by the obvious differences observable in members of the same family. It is moreover far from obvious, in considering the causes which might influence a young girl's ideas on this subject, that the environment of a large family would favour a higher degree of fertility than that of a small one. The example of a mother must act both by suggestion and by counter suggestion, and it would seem hazardous to assume that girls are more strongly influenced, in practice, by a consciousness of the special advantages of their home environment, than by a consciousness of its special disabilities.

However this may be, we possess in the relationship between the fertility of granddaughters and that of their paternal grandmothers, an unequivocal confirmation of the view that the relationship observed between mother and daughter is essentially one of organic inheritance. For a corresponding table shows that the regression of the granddaughter's fertility on that of the grandmother's is 0·1065, almost exactly half of that found for the influence of the mother. An effect due wholly to organic inheritance should approximately be halved in each generation, apart from any strong tendency towards homogamy, and although the grandparental effect especially, though statistically significant, is subject to somewhat large sampling errors, its coincidence with the expected value undoubtedly points to organic inheritance as responsible, at least, for the greater part of the correlations observed.

It is instructive to put this agreement in another way. In the table between mother and daughter, the covariance[1] between the two fertilities is 20·9 per cent. of the variance observed in the fertility of the daughters. In characters almost wholly determined by inheritance, the covariance should be almost half of the variance, and we may thus make the estimate that 41·8 per cent. of the observed variance in fertility is ascribable to heritable causes. In Chapter II it was pointed out that a good estimate of the genetic element in the variance of any character could be obtained by dividing the square of the correlation between parent and child, by the correlation between grandparent and grandchild. In the present instance this gives us $(0·2096)^2 \div 0·1123$ or 39·1 per cent.   Both methods of calculation therefore agree in ascribing about 40 per cent. of the observed variance to heritable causes, and this estimate, when we consider the enormous predominance of the random element in individual survival obtained in the discussion in Chapter IV, shows that the heritable factors influencing human fertility must exert a selective effect of quite exceptional magnitude.

We may express the selective intensities thus indicated in the units we have used hitherto to express differences in the Malthusian parameters, namely per cent. per generation, by recognizing that the genetic variance arrived at in the preceding paragraph is in actual magnitude 3·042 units, corresponding to a standard deviation, due to hereditary causes only, of 1·745 children per family. Since the average number of children born to the daughters is 3·923, individuals or stocks differing from the average by no more than the standard deviation, in opposite directions, will have families in the ratio 5·668 : 2·178. If the generations are of equal length, we may obtain

---

[1] The mean product of the deviations from the means.

the selective advantage of the more fertile over the less fertile groups, by finding the difference between the natural logarithms of these two numbers. This difference is 0·9564, indicating that the selective advantage produced by variations (of by no means exceptional magnitude) in innate fertility, amounts to over 95 per cent. in each generation. In this form the statement of the intensity of selection is comparable to the selection of the order of one per cent. which it has hitherto been found convenient to consider in exemplifying the numerical effects of selections of determinate intensity.

The evolution of the conscience respecting voluntary reproduction

The calculations of the last section indicate that among our contemporaries, at least, differences in heritable disposition play so large a part in the determination of fertility, that somewhat large modifications of temperament must be brought about by reproductive selection within a span of ten generations, that is within a relatively short historical period. We have seen that by no means extreme genotypes have rates of reproduction in a ratio higher than 2 to 1, and if for simplicity we imagine a population consisting of two strains, which, having equal mortality, differ only to this extent in fertility, it is easy to see that the numbers of one type relative to that of the other will, in ten generations, be increased over a thousandfold. To put the matter in another light, if at the beginning of the period the population consisted of 97 per cent. of the less fertile and 3 per cent. of the more fertile strain, in 5 generations the two strains are brought to an equality, and in 5 more their situations are reversed, so that the less fertile strain is represented by 3 and the more fertile by 97 per cent. of the population. Civilized man, in fact, judging by the fertility statistics of our own time, is apparently subjected to a selective process of an intensity approaching a hundredfold the intensities we can expect to find among wild animals, with the possible exception of groups which have suffered a recent and profound change in their environment. We may therefore anticipate that a correspondingly rapid evolution has taken place within historical times in the appropriate mental attributes.

As is well known, the practice of infanticide is somewhat widespread among uncivilized peoples. If we consider the perils of savage life, and the extreme hardships to which uncivilized peoples are from time to time exposed, it would, I believe, be a mistake to regard this practice as altogether a maladaptation. Wherever the mortality in infancy and childhood is extremely high, the reproductive value, in the sense of Chapter II, of a new-born infant, must be small compared to that of a young and fertile woman. In times of famine, or of urgently enforced migration, an attempt to spare the life of the child would not only be often unsuccessful, but would certainly endanger the more valuable life of its mother. The act of infanticide, however offensive to civilized feelings, has, in these circumstances, a certain natural propriety, and it is not surprising that it should often be regarded as a moral action and be, to some extent, compulsory. The motive may be regarded as economic; but since among peoples with whom the accumulation of property is at a minimum, foresight as to future resources is only felt under urgent stress, their motives can scarcely be called prudential. Among settled peoples, on the contrary, prudence and the accumulation of property begin to play an essential role, while the immediate effect of increasing prosperity, by diminishing the selective importance of the death-rate, is to enhance that of the rate of reproduction. It is not surprising therefore that the religions of all civilized peoples, without exception, should expressly condemn infanticide, although in several cases, such as the peoples of Northern Europe and the pre-Islamic Arabs, we know it to have been the practice of their barbarous ancestors.

It is evident that the natural instincts of the father and mother must offer some resistance to the practice of infanticide, even when it is performed under the pressure of need. Moreover, we cannot doubt that savages, like civilized people, differ among themselves in their degree of callousness or sensibility. The feeling of aversion and repugnance to the cruel act must be more strongly developed in some parents than in others, and since infanticide is always to some extent at the discretion of the parents, it cannot be doubted that the former allow a larger number of offspring to live. In this way there is a natural tendency, wherever infanticide is practised without the pressure of the severest hardship, to strengthen the feelings of tenderness and compassion towards the newly born child, by the natural elimination of those who are the most willing to murder their offspring, for the sake of an easier or a freer life.

Amongst wholly savage peoples, although infanticide is not usually regarded as wrong, it is probably not practised more frequently than is on the whole racially advantageous. If however, in the course of time, a people, with feelings in this matter appropriate to the condition of extreme savagery, came to be placed in conditions in which the accumulation of property is not only possible, but is the natural aim of the more ambitious, the temptation to infanticide will be more uniformly insistent, and the corresponding selection of the moral instincts which resist this temptation, will be correspondingly severe. If we accept the view that all long civilized peoples have been purged of their more callous or murderous elements by passing through this period of severe selection, we shall be in a position to realize why it is that the consciences of civilized peoples, as expressed in their religious teaching, should so unhesitatingly condemn infanticide.

An interesting example of such a transitional period is supplied by the Arabs. The extreme hardship of the life of the desert Bedouin must have presented many occasions in which infanticide, was, not merely economically, but biologically advantageous. Property existed, but was principally tribal, and tribal feeling is evident in the restriction of infanticide to female children. In the pre-Islamic period we hear of reluctant parents being urged to kill their baby daughters for the sake of their tribe. In the small trading centres the conditions of life were doubtless much ameliorated, and the possession of property was certainly largely individualized, though moral opinion was doubtless dominated by the Bedouin tradition. The Koraish, the tribe of Muhammad, seems to have been in possession of Mecca for about five generations before his birth. At that time it is evident that the public conscience was beginning to rebel against the usages of their ancestors, from the importance which his little band of proselytes attached to abstaining from infanticide. It is abjured in one of the few clauses of the 'First Homage', and is repeatedly forbidden in the Koran. One of the first acts of Muhammad, after his conquest of Mecca, was to obtain an oath of abstinence from infanticide from the Meccan women. To avoid the misapprehension that this doctrine was connected with the military ambitions of the Moslem state, it should be noted that the 'First Homage' was taken before Muhammad's flight from Mecca, at a time when the idea of upholding their faith by force of arms was entirely remote from the minds of the faithful.

If it be conceded that the selective effect of infanticide has been gradually to extinguish those hereditary types of temperament which are least unwilling to conform to this savage custom, a similar selective effect is to be expected from the equivalent act of feticide or abortion, whenever from prudential motives this practice has become sufficiently common. The history of the evolution of moral opinion in ancient Greece and Rome, during a period when feticide was very widely practised, is therefore of considerable interest. It is fortunate that an examination of the historical evidence relevant to the theories under discussion does not require a detailed knowledge of the practice of peoples at different epochs, but of the prevailing moral opinion as to what is right and what is reprehensible. An investigation into practice would be a matter of the greatest difficulty, as is evident from our ignorance of the current practices of our own time. The practice of different individuals at any one time is, moreover, certainly far more variable than their moral opinions, which are usually held in common to a rather remarkable degree. The majority of those who do what is thought to be wrong certainly do so without departing far, in their moral opinions, from the rest of the society to which they belong. We can seldom indeed find distinct differences of moral opinion, even between the protagonists in religious controversy. If, therefore, we find pronouncements upon social practices in the writings of moral philosophers of high repute in their own generation, we are on very safe ground in inferring that their expressed opinions on these subjects were not shocking to any considerable body of their contemporaries, but, on the contrary, were nicely designed to express the common ground of thoughtful and high-minded opinion in their own time.

In the fourth and fifth centuries B.C. no opprobium seems to have been attached to the practice of feticide. We find high-minded men of the greatest moral weight commending its utility. Plato advocates it principally for eugenic, Aristotle for economic motives, much as high-minded men of our own day may be found to advocate more modern methods of birth limitation. A few centuries later a somewhat changed outlook is unmistakable. Polybius, in whose time, in the second century B.C., the increase of the custom had led to an evident depopulation, regards this as a political evil, and evidently felt that the rich, at least, should be blamed as unpatriotic. 'For this evil grew upon us rapidly, and without attracting attention, by our men becoming perverted to a passion for show and money and the pleasures of an idle life, and accordingly either not marrying at all, or, if they did marry, refusing to rear the children that were born, or at most one or two out of a number, for the sake of leaving them well off or bringing them up in extravagant luxury.' He believes that this should be changed, 'by the men themselves if possible changing their objects of ambition; or, if that cannot be done, by passing laws for the preservation of infants'. In the first century A.D. Pliny, whom we have already quoted, is almost apologetic, but evidently broadminded. Seneca is more severe. Writing in praise of his mother Helvia he specially mentions her virtue in never having committed feticide, an act which he evidently reckons among the follies and vices of the fashionable world. We may see by these instances, and by many other less direct indications, that a profound change had taken place in the public conscience of the Roman Empire before its conversion to Christianity. A few centuries later, what had seemed patriotic and high-minded to Plato and Aristotle, a venial offence and a fashionable folly to Pliny and Seneca, was anathematized as a damnable sin by the Christian fathers, such as Tertullian, and was made a capital offence in the laws of Valentinian.

Since the change in moral opinion with respect to feticide, together with associated changes in sexual morality, took place over a period during which the ancient world experienced the religious revolution of conversion to Christianity, and since these concurrent changes were ultimately connected causally, it is important to avoid the misapprehension that the change in moral feeling, with which we are concerned, was derived as a consequence from the acceptance of Christian doctrine. It would, I believe, be a fundamental mistake to imagine that the moral attitude of any religious community is to any important extent deducible from the intellectual conceptions of their theology (however much preachers make it their business so to deduce it), and still more to suppose that official doctrine is not itself largely moulded by the state of the popular conscience. The facts available to guide us in the present case are:

(i) The progressive change of moral opinion among pagan writers from Plato to the jurist Julius Paulus, who, about the beginning of the third century A.D., stated the opinion that the exposure of children was morally, though not legally, equivalent to murder, and expressed strong disapproval of the practice of feticide. The opinion of these writers, who either knew nothing of Christianity, or rejected its doctrine, cannot easily be ascribed to the state of Christian feeling upon the subject.

(ii) Our earliest authentic source of Christian doctrine, the story as told in the Gospels, teaches an attitude of gentleness, forbearance and self-sacrifice which has ever since been regarded as-ideally Christian. It contains, however, no condemnation of feticide; and cannot even be regarded as the source of the theological doctrine of the damnation of unbaptized infants, even if born dead, which undoubtedly at a later date exerted an important influence in crystallizing Christian opinion upon this subject. It cannot easily be argued that, while not specifically condemning feticide, the gentle and humane character of the Gospel teaching in reality determined the later attitude of the church, for there is nothing gentle or humane in the doctrine we have quoted, and the early Christian legislation against sexual offences certainly seems to err on the side of excessive ferocity. On a statute of Constantine Gibbon remarks:

> The successful ravisher was punished with death; and as if simple death were inadequate to the enormity of his guilt he was either burnt alive, or torn in pieces by wild beasts in the amphitheatre. The virgin's declaration that she had been carried away with her own consent, instead of saving her lover, exposed her to share his fate. The slaves, whether male or female, who were convicted of having been accessory to the rape or seduction, were burnt alive or put to death by the ingenious torture of pouring down their throats a quantity of melted lead.

We can understand the temper of such legislation as the work of men filled with a temperamental fear and loathing of sexual licence, working in a frenzy of fanatical enthusiasm; we cannot explain it as the work of broad-minded and tolerant men convinced by the Gospel story of the divinity of Jesus.

(iii) That a stricter sexual morality, including a more vehement condemnation of feticide, did in fact characterize Christian as opposed to Pagan ethics, when these are contrasted contemporaneously, must be ascribed in large part to the direct appeal made by Christianity to the sanction of the individual conscience, which made it the more responsive to popular feeling, even when latent and unexpressed. This effect was certainly enhanced by the solicitude of the early Church for persons of the meanest occupational status, since, as will be shown more fully in Chapter XI, the evolutionary effects of variations in reproduction are greatest in the lowest social class. In these respects the influence of the rise of Christianity upon the change in moral opinion was intimate and direct; it would be beyond the scope of our subject to consider this association as a factor in determining the triumph of Christianity in the Roman world.

Analogies of animal instinct and immunity to disease

To a remote spectator the evolution which the human instincts of reproduction have undergone during civilized periods would seem scarcely more remarkable in kind than the analogous modifications which the reproductive instincts of other animals have undoubtedly experienced in their evolutionary development. The principal contrast would probably seem to lie in the great intensity of the selective process in Man, and the correspondingly rapid evolutionary progress, in spite of the great length of the human generation. The moral qualities are so important to ourselves, and appear to possess so absolute a sanction, that it is not easy at once to take up this remote view-point; yet any sympathetic consideration of the contingencies of reproduction among the higher animals must convince us that their moral nature also must have been in many cases modified by a strictly analogous form of selection. The astonishing courage shown by many ordinarily timid species of mammals and birds in the defence of their young, and even the abstention of carnivorous mothers from infanticide, are cases in which we can at least be sure that the emotional response to a situation connected with reproduction has been modified in a manner analogous to the modifications of the moral emotions found necessary in man to overcome the temptation to the specific vices to which he is exposed.

An even closer analogy might perhaps be found in the evolution of immunity to specific diseases and to morbid cravings as for alcohol, through the selective action of the death-rate, parallel to the evolution of immunity to sexual vices through the selective action of the rate of reproduction. In the case of diseases it is, I believe, accepted that a special or specific immunity must be acquired independently against each disease in turn; on the other hand it is I think evident that immunity to such crude vices as infanticide is usually acquired through mental modifications, which themselves aid in conferring some degree of immunity to the more insidious temptations to feticide or to contraception. Certain types of vital statistics are at least suggestive of the view that, if they could be tested under equal conditions in the same environment, the most anciently civilized peoples, or rather those groups in whose collective ancestry exposure to civilized conditions has been greatest, are the most prolific, while those most recently civilized are least able to maintain their numbers in a civilized community. For example, in the maternity statistics of the State of Rhode Island as reported by Dr. Hoffmann we have the proportion of women married at the time of the Census (1905) who at that time were still childless, stated according to religious denominations, Protestant, Catholic, or Jew, and for American born and Foreign born women respectively. The percentages given are:

TABLE 10.

	American born.	Foreign born.
Protestant . .	30·7	19·4
Catholic . .	24·2	16·9
Jewish . . .	18·9	11·4

The contrast between the American born and the Foreign born is apparently one of social class comparable to the occupational differences to be considered more fully in the next chapter. The religious differences might conceivably be ascribed to the direct effect of religious doctrine as to the practice of contraception, which is condemned by the religious heads of both the Catholic and of the Jewish communities. It seems more probable that the differences are racial in origin, and on this view it is at least interesting that the racial stock of Protestant peoples should be relatively recently civilized, while that of the Jews has been largely exposed to the influence of civilization from times prior even to the classical civilization of the Mediterranean area.

The bare fact of a people having become gradually averse to a specific action, such as feticide, owing to its effect upon the proportionate survival of different individuals in the ancestry of subsequent generations, throws very little light upon the mental processes by which this feeling of aversion is brought about. As a cause of such changes Natural Selection is entirely indifferent to the means, so long as the required end of effective reproduction is achieved. The Chinese belief respecting infanticide, that the ghosts of the murdered babies bring misfortune, indicates a method of modifying the reproductive instincts by the preferential survival of the more credulous and superstitiously fearful. It may well be considered that the source of such a belief is to be found in an innate tendency to experience a feeling of guilt on committing infanticide; but when such a belief has grown up, and has become widely accepted, its existence will undoubtedly modify the incidence of selection in the particular mental qualities which are favoured. Although the progressive strengthening of the reproductive instincts, which may be regarded as the principal effect of reproductive selection, will be steadily pursued, a variety of psychological modifications, together with sociological changes consequent upon these, may all be, in different circumstances, effective means to that end. It is convenient to regard as the primary effect of selection directed against a conscious and voluntary action, the development of moral aversion towards that action, while other mental attributes which produce the same effect, through the pursuit of illusory ends, constitute a kind of mimicry of the appropriate morality. Which type of mental attribute will turn out in practice to be the most effective, and in consequence be subject to the most rapid modification, must depend in part on the extent to which knowledge of natural causation is diffused among the people in question.

A consideration which is relevant to the conscious choice of the parents in reproduction, equally with their moral attitude towards the methods of family limitation available, is the motive for which that limitation is desired. Parents in whom economic ambition is strong, will, in like circumstances, be more inclined to limit their families than those in whom it is weak. Consequently a progressive weakening of the economic ambition, or at least in the average intensity with which this motive is felt among the great body of citizens, is to be expected as a concomitant to the strengthening of the moral aversion towards family limitation. The attitude of men and women towards their economic welfare cannot, however, be ordinarily reduced by this cause to indifference, for in countries in which the poorest class are frequently decimated by famine, it is apparent that a stage will be reached at which what is gained in the birth-rate is lost in the death-rate; and even where the extremes of distress are ordinarily avoided, some loss of civil liberty, and of the opportunities for reproduction, has been the common effect of indigence. Moreover the rational pursuit of economic advantage must, even in the most civilized countries, frequently place the individual in a position favourable to normal reproduction. It would, apparently, in most societies, be as disastrous to the biological prospects of the individual to lack entirely the acquisitive instincts as to lack the primary impulses of sex, notwithstanding that the abuse of either passion must meet with counter-selection. The moral attitude of civilized man towards money, as towards sex, must be therefore the product of much more complicated evolutionary forces than is his attitude towards infanticide or feticide, as might perhaps be inferred from the hypocrisy and fanaticism, the passions and the passionate inhibitions, found among long-civilized peoples on both subjects. We may at least gather a hint of one of the reasons for the moral disrepute commonly associated with mercenary motives, a disrepute which itself contrasts rather strangely with the economic theory that payments are the conventional equivalents of services freely recognized as worth such payment; though there is nothing in our argument so far to show why the prevailing attitude towards wealth should differ from that sketched at the end of the last chapter, where it was assumed that the normal destiny of accumulated wealth was to provide for a numerous posterity.

Summary

The number of children actually born to different individuals in civilized societies is, as in other organisms, largely influenced by chance. Whereas in most wild organisms the contribution of other causes to the actual variation is probably so small that it could not easily be detected, the total variance in the number of offspring produced in civilized man is so great that a considerable fraction must be ascribed to causes other than chance.

Temperamental qualities exert a great influence in determining celibacy, or age at marriage, both in men and women. They are, and throughout human history generally have been, at least equally important in conditioning the use or abstinence from the use of artificial methods of family limitation.

The number of children born to women is significantly correlated with the number born to their mothers. Statistics of the upper social classes in the nineteenth century suggest that about 40 per cent. of the total variance observed may be ascribed to heritable causes. Of these by far the most important must be inherited qualities of the mind, and especially of the moral temperament. The intensity of selection by differences of fertility, due to innate causes of the order found, is relatively enormous in comparison to selective intensities to be expected in nature. In spite of the late maturity of man, it is sufficient to produce considerable evolutionary changes in relatively short historical periods.

The custom of infanticide in the towns of pre-Islamic Arabia was accompanied by a change in moral outlook which finally forbade the custom. The widespread use of exposure and feticide in the earlier ages of the Graeco-Roman civilization was accompanied by a gradual and persistent change of opinion in regard to these practices, until they finally became capital crimes. The influence of Christian teaching upon the abolition of these views was very great, but this influence sprang rather from the consciences of the people than from dogmatic doctrine.

Races may be regarded as becoming immune to race-destroying vices in a manner analogous to the acquirement of immunity against disease by means of selective mortality. There is some evidence in modern statistics that those races are most immune, even to modern methods of birth limitation, which have had the longest social experience of civilized conditions.

The evidences examined leave little room for doubt that the most powerful selective agency in civilized man is that acting upon the mental and moral qualities by way of the birth-rate, and that this agency is at work with an intensity which cannot easily be paralleled in other species.

X

REPRODUCTION IN RELATION TO SOCIAL CLASS

Economic and biological aspects of class distinctions. Defects of current data. Early investigations. British data. Position in the U.S.A. Effects of differential fertility. Summary.

The world will be thy widow and still weep
That thou no form of thee hast left behind. SHAKESPEARE.

Economic and biological aspects of class distinctions

FOR the biological study of mankind the institution of social class has a double importance. Merely from the economic standpoint the different occupations are graded in the rewards which they secure to their professors, and, it is to be presumed, in the social value of the services performed. From this point of view, the social class which a family occupies may be regarded as the margin of prosperity which separates it from destitution, or at least from a level of bare subsistence; so that a comparative failure in one occupation is free, at the expense of some loss of social class, to find some humbler mode of living in which his talents are sufficient to win the necessary support for himself and his family. The elasticity afforded by the range of occupations, and by the differences in the rewards which they procure, may thus be regarded as mitigating the severity of the consequences which failure entails upon solitary organisms, by leaving for the majority of the population a certain margin of safety from absolute destitution.

An aspect of social class of more direct biological importance lies in the profound influence which it exerts upon the intermarriage of different families. In several instances the importance of avoiding class intermarriage has been so strongly felt that such marriages have been opposed by definite legal penalties, with the consequent formation of closed, more or less aristocratic, castes. This has been especially the case where racial differences have been involved. Without any such legal provision, however, prevailing opinion, mutual interest, and the opportunities for social intercourse, have proved themselves sufficient, in all civilized societies, to lay on the great majority of marriages the restriction that the parties shall be of approximately equal social class. In this statement social class should, of course, be taken to comprehend, not merely income or wealth, but also the prestige attaching to occupation, personal talents, and family associations. Its meaning is thus somewhat different from, though closely correlated with, the purely economic use of the term. But the factor of intermarriage is so important in its social and biological consequences that it will be best to use the term 'social class' solely in this sense and to lay down that the social class of an individual or his family shall be defined by the aggregate of persons or families, intermarriage with whom will encounter no social obstacles.

By reason of the influence of class on mate selection, distinctions of social class are distinctions of relatively permanent biological entities, not indeed strictly insulated, but influencing one another by a constant diffusion of persons who pass from class to class. This mutual diffusion should be distinguished from the great tide of social promotion, which, as will be seen more clearly, is now everywhere in progress, and upon which the mutual influence by diffusion may be regarded as superimposed. These two social movements are, of course, confused in the history of each several individual, but their causes and consequences must clearly be distinguished. The relatively permanent character of the social classes gives a special importance to the differences which exist in their vital statistics, for, to produce an evolutionary change it is necessary that groups contrasted, for example, in their rates of reproduction, should also be differentiated genetically; and if this change is to be progressive and continuous it is further necessary that like differences in reproduction should continue generation after generation to characterize groups showing the same genetic contrasts.

That a puzzling discrepancy existed between the birth-rates of certain highly placed families, and those of the general population, seems to have been suspected during a great part of the nineteenth century; but neither the magnitude nor the geographical extent, nor, indeed, the sociological nature of the phenomenon seems to have been appreciated until quite recent times. The theory, of unknown origin, that the 'inbreeding' supposed to be practised by exclusive aristocracies, was a cause of sterility (and of other signs of degeneracy) was doubtless framed under the impression that wealthy plebeian families were as fertile as the rest of the community. Herbert Spencer's theory of an inherent opposition of 'individuation' and 'genesis' would seem to suggest that he recognized some failure of reproduction as characteristic of the brain-workers. One of the most important points established by the modern data is that the deficiency in procreation is not specially characteristic of titled families, or of the higher intellects, but is a graded quality extending by a regular declivity from the top to the bottom of the social scale.

Defects of current data

Before considering some of the more important researches in this field it is necessary to point out that, although the general character of the facts may be said to be established, in most western countries with certainty, yet the actual quantitative data available are far less satisfactory than they might easily be made, and this for two principal reasons. In the first place, they are much out of date, and relate, in some of the most important instances, to the rate at which births were occurring in different classes as much as thirty-five years ago; while as to what is now occurring we can only conjecture from the past course of events. The best data for England and Scotland as a whole, for example, are still those which were obtained in the census of 1911, from the completed families of persons enumerated in that census. Many of the children upon which these comparisons rest were, therefore, born quite as early as 1890 and represent the behaviour rather of the last than of the present generation of parents. Such an interval of time may be of great importance in bringing to pass changes in the birth-rate; for it is certain that, whatever changes have occurred in the relative rates of reproduction of different classes, the birth-rate of the country as a whole has diminished very greatly during this period. Although the questions may be repeated in the census of 1931 it is clear that comparisons based upon statements of completed families will, by the time they are made public, be already nearly twenty years out of date. Direct information as to the course of natality from year to year could, however, be obtained from the numbers of births registered, provided the ages of the parents at such times were recorded, and the occupation of the father, which is now recorded, were brought into harmony with the occupational classification employed in the census.

In the second place, a completely satisfactory comparison between different occupational groups must involve the mortalities occurring in those groups, and the natalities as affected, not only by the birth-rates of married persons of a given age, but also by age at marriage and frequency of celibacy. The basis for a complete biological comparison is, in fact, to be found only in the Malthusian parameter of population growth, of which the theory was developed in Chapter II ; it is therefore only when the Malthusian parameters for different occupational groups have been established, that a quantitative knowledge of the biological situation will have been obtained. In all the examples to be given it will be seen that our knowledge falls very far short of this level. It is perhaps of minor importance that mortality, except in early childhood, should have been generally ignored, for this is certainly not the major factor of the comparison. Even with respect to births, however, nearly all inquiries tend to confine themselves to a comparison of the births to married persons of a fixed age, and often assemble their data in such a way as to eliminate, as far as possible, the important factors of age at marriage and celibacy. It is, of course, desirable in itself, that these three elements should be evaluated separately, and there are reasons to think that their separate evaluation will be of service in disentangling the causes of the phenomenon. Yet, since no one item is of the same practical significance as the total effect, more is usually lost than gained by forms of tabulation in which the total effect as such is ignored.

Early investigations

One of the earliest important investigations was that carried out by Dr. David Heron upon the birth-rates in the different metropolitan boroughs into which London is divided. His method consists in the correlation of fertility with a number of indices of social status, derived from the proportion of different occupations recorded in the census. The districts in which the average social level is highest are indicated by a high proportion of professional men and of domestic servants, and by a low proportion of general labourers, pawnbrokers, employed children, and occupants averaging more than two in a room. The fertility was measured in 1901 by the birth-rate for a thousand married women aged 15 to 54, and in 1851 by the birth-rate for a thousand married women over 20. The birth-rate was found to be closely associated with all indications of a lower social class. Both epochs show the same general relationships between undesirable social conditions and a high birth-rate ; the intensity of this relationship, however, has almost doubled in the interval from 1851 to 1901. The investigation is of particular interest in showing that in 1851 a class difference in fertility had already developed, although of only half the intensity of that shown at the end of the century. The comparison of intensity is, however, obscured by the change which Dr. Heron was obliged to make in the method of measuring fertility, for the slightly inferior index used for 1851 would itself probably diminish the intensity of the correlations obtained. It should be remembered that the comparison is made only between districts, and that what is demonstrated is no more than that, in London, the poorest districts have the highest birth-rates. It does not prove that the poorest classes are themselves responsible for these birth-rates, although such a general association is a natural inference from the results. The methods used, however, cannot be made to distinguish whether the cause of the contrast is that the better paid workman is less prolific than the worse paid, or merely that the professional and salaried occupations are less prolific than the working classes in general. The differences observed in this study do not of course include any differences in the incidence of celibacy, or in the average duration of marriage before 45.

The first national statistics in which size of family can be compared with occupational status were obtained in the French census of 1906. For families of which the head was 60 to 65 years old at the time of that census, and still married, the average family was 3·60. For employers (*patrons*) the average was 3·59, or almost equal to that of the general population; for workmen (*ouvriers*) it was 4·04, while among *employés* other than workmen it was only 3·00. The *patrons* are more than half farmers, for whom the average for those married more than 25 years is 3·71, while little more than a fifth of them belong to the liberal professions, for whom the corresponding average is 3·30, or a trifle less than that of the *employés*, other than workmen, if allowance is made for the change in the basis of classification.

The classification of the census was, however, better fitted to compare the conditions in different industries than to compare those in different grades of the same industry. Fresh evidence was obtained in 1907 from schedules of families filled in by a large number of employees and workmen, in the pay of the state, and of various provinces and parishes. The averages for marriages which had lasted more than fifteen years show for workmen a regular decrease in size of family from the worst paid, with an average of 3·29, to the best paid, earning from £160 to £240 a year, with an average of 2·34. The employees, in the same wage-groups show a corresponding fall from

2·77 to 2·31. Those earning from £240 to £400 a year show a further fall to 2·29, while those earning more than £400 a year show a rise to 2·38. It is clear from these and the preceding figures, that in the bulk of the French population, earning less than £400 a year, reproduction, at a period that must be referred back at least to the eighties and nineties of the last century, was much more active among the poorer than among the better-paid classes. For officials earning more than £400 a year, at which point the salary subdivisions cease, there was, apparently a slight rise in reproduction, although within the more prosperous classes the intellectual group represented by the liberal professions had much fewer children than the average.

For Great Britain the fullest information so far made available is that obtained from the census of 1911, but a number of private investigations had already made clear individually the facts of the greatest importance. Mr. Whetham in 1909 published the results of a study (*Family and Nation,* p. 140) of the fertility of the higher ranks of the professional and official classes, in which he made use of the information supplied by *Who's Who.* He finds that the average number of children, recorded for married persons in this class, falls from 5·19 to 3·08, in passing from those married before to those married after 1870. He notes that for clergymen the diminution is much less, and for military families considerably more, than for the remainder. A similar investigation by the same author shows that, for families that have held a peerage for at least three generations, the average has fallen, according to the date of marriage, from 7·10 for marriages in the thirties of the last century, to 3·13 for those contracted in the eighties. Although these investigations are confined to a small fraction of the British people they reveal significant facts, first as to the low rate of reproduction of the English upper classes of the last generation, and secondly that this rate of reproduction was falling during the latter half of the nineteenth century at a rate greatly exceeding the decline in the fertility of the general population.

It is sometimes imagined that the differences in reproduction between different social classes, depend upon circumstances and ideas peculiar to those who enjoy the advantages of leisure, or education, or both combined. It seems frequently to have been deemed probable that where poverty is severe the most healthy and efficient will reproduce their kind the most freely, as is the case with animals in a state of nature. Whatever may have been the case in the past, a careful and most valuable investigation has shown that data, drawn from the poorest classes in the industrial towns of Blackburn, Preston, Glasgow, and Birmingham, and from the Royal Albert Asylum, Lancaster, prove that the most capable among these classes have the fewest children. The correlations found are not large, as is to be expected when we deal with variations within a population of nearly uniform culture. What is established beyond doubt is that, when proper allowance is made for the age of father and mother the larger families subsist upon the smaller wages. In other words the poorer wage earners have, at a given age, the greater number of surviving children. Similar results are obtained when other tests of capacity or status, such as health, regularity of employment, temperance of the father, or cleanliness of the home, are substituted for wages. Of these criteria the least satisfactory is the cleanliness of the home, since it is probable that visitors, in grading this variate, do not make accurate allowance for the extra work entailed by the larger families.

British data

The results of the census of 1911 have been reported upon separately by Dr. J. C. Dunlop for Scotland and Dr. T. H. C. Stevenson for England. The two accounts differ considerably in the form in which the results are reported, and since the results of the two inquiries are closely similar, each serves in several valuable respects to supplement the other. For Scotland Dr. Dunlop gives a comparative table of the average numbers of children for a large number of occupations. Only marriages which had lasted fifteen years or more were utilized, and the age at marriage of the wife was restricted to the five year period 22 to 26. The average for all occupations is 5·82 children, and by arranging the occupations in order of the average size of family it is easy to compare the social status of the most fertile, with that of the least fertile occupations. To avoid giving undue weight to the small groups, for which the average fertility will not be ·very precisely determined, we should also take account of the number of families represented by each average. The most fertile occupations comprise three large groups, namely, crofters 7·04, coal, shale and ironstone miners 7·01, agricultural labourers and farm-servants 6·42; three groups of moderate size, namely old-age pensioners 6·95, fisherman-crofters 6·93, and coal heavers 6·61, besides the small group of plasterers' labourers 7·01. Turning now to the least fertile occupations the only large group is the very composite one of clerks, scoring 4.38. Six groups of moderate size are the physicians and surgeons 3·91, advocates and solicitors 3·92, literary and scientific 4·09, schoolmasters and teachers 4·25, art, music and drama 4·27, ministers and clergymen 4·33. In addition there are three small groups of army officers 3·76, dentists and assistants 3·86, and veterinary surgeons 4·00. The contrast between the social status of these two sets of occupations requires no stress. Among the least fertile are the learned professions, and, among the most fertile, the poorest of manual labourers. The differences in the average numbers of children are the more remarkable owing to the severe restriction to which the age of the mother has been subjected. Together with the restriction on the duration of marriage, this is nearly equivalent to eliminating, between the different occupational classes, all such differences in fertility as arise from differences in the frequency of celibacy, and in the age of marriage. It clearly appears from the corresponding English figures that the contrast between the real rates of reproduction in the different classes, is heightened by the inclusion of these additional factors.

The census results are particularly valuable in showing that the decrease of fertility with improved social class is not confined to any one section of the social scale, as it seemed to be in the French figures already quoted, but is equally manifest from the very poorest groups that can be compared, to the most distinguished occupations which provide sufficient numbers on which to base a comparison.

In several instances it is, moreover, possible to compare the fertility of groups which are essentially grades within the same occupations. The comparison of these is of particular importance, because we thereby eliminate the considerable factors of differences in locality and of industrial environment, and in the absence of these, the averages derived, even from the smaller occupational categories, possess considerable significance and regularity. Thus the three pairs of categories in Dr. Dunlop's list, which seem to be strictly comparable, show us that bricklayers' labourers have more children than bricklayers, that agricultural labourers have more children than farm foremen, and that plasterers' labourers have more children than plasterers. Since in each of these comparisons all variations in locality, and in the conditions of employment, must be shared, in a somewhat strict proportion, between the two grades of labour, these comparisons display the fertility differences in the clearest light. Since 1911 the classification by occupations, used in the census, has been very greatly improved, especially by separating the classification by industries from the classification by grade or employment within each industry, and if the inquiry as to number of children is repeated in the census of 1931, many more comparisons between different grades of workers in the same industry should become possible.

In 1920 Dr. Stevenson published an account of the results for England and Wales in which special attention was given to overcoming the difficulties of obtaining conclusions of direct sociological value from data, the collection of which had been, for that purpose, very imperfectly planned. The occupational categories, although they include distinctions both of industry and of occupation proper, could be roughly graded into five large sections, of which the first represented the upper and middle classes, in a very generous interpretation of this term, the third was of skilled workers and the fifth of unskilled. Owing to the absence of grade distinctions it was necessary to omit from this subdivision the three great groups of textile workers, coal miners, and agricultural labourers.

For the families of continuing fertility, i. e. where the wife was under the age of 45 years at the date of census, the average numbers of living children for these five groups are 1·68, 2·05, 2·32, 2·37, and 2·68, the unskilled workers having on the average rather less than twice the number of children found in class I. The greatest contrasts are between classes I and II, with a difference 0·37, and classes IV and V, with a difference 0·31, showing decided decreases in fertility associated with the lower, as well as the higher, grades of ability. Standardization with respect to age at marriage and duration of marriage reduces the differences by about a third, thus showing that the classes most fertile when married at any given age, also provide for themselves by early marriage the greatest opportunity for fertility. It also appears from the census report that the proportion married is less in the professional classes than in the remainder of the population. Consequently the disparity in reproduction receives contributions from three distinct sources. In the first place the upper classes marry less, or remain celibate more frequently; secondly they marry later, and their wives during marriage are for a shorter time in the reproductive period, and that in the later and less fertile portions of that period; thirdly, the number of children recorded is fewer, compared to the number of wives of any given age. The third cause is probably of recent growth, and may principally be ascribed to the deliberate limitation of births ; but it should be noted that this explanation is not available for the important contribution of the first two causes, which would tend to be diminished rather than increased by the increasing use of contraception.

Dr. Stevenson attempts to trace the earlier history of the relative fertility of different classes by comparing the average families, in the different occupational groups, separately for marriages contracted in successive decades, back to 1851–61, from which marriages a sufficient number of husbands survived to the 1911 census. The occupational contrast in the fertilities of these early marriages is found to be considerably reduced, and Dr. Stevenson suggests that, if the comparison could have been carried twenty years further back, a period of substantial equality between all classes might have been met with. The data from a single census, which are all that we possess, are, however, quite inadequate to supply the necessary information on this subject; for the marriages on which the census averages are based are not a random sample of those which would have been recorded at an earlier census, but a sample strictly selected on the basis of longevity. In pedigree data of the middle class it is common to find appreciable correlation between fertility and longevity, and the validity of the earlier figures must be substantially affected from this cause. The argument from convergence not only requires that the occupational classification should be equally reliable for the oldest groups, as for those still in employment, but that the effect, upon the fertility of the survivors, of selection for longevity should be effectively the same in all classes. From a series of comparable census investigations some such extrapolation should be feasible, but since the opportunity was lost in 1921, the class differences in reproduction, before the general fall of the birth-rate set in about 1875, can now scarcely be recovered. It is clear, however, from Heron's work, which we have quoted above, that substantial differences existed between the different London boroughs prior to 1851. While the period of class equality is thus quite hypothetical, the increase in the disparity, since the middle of the nineteenth century cannot reasonably be doubted.

The general decline in the birth-rate since the seventies has affected all classes, though somewhat unequally. The decline has been greater among the more prosperous, and less among the poorer, classes. There seems to be no statistical justification for the assertion that the decline has 'spread from above downwards', nor is this inherently probable, since propaganda among the poorest classes has been characteristic of the neo-Malthusian movement from its inception. The point is to be emphasized, since it appears to be somewhat widely believed that the differential birth-rate, in so far as it is due to deliberate family limitation, will automatically disappear, by the simple process of the poorer classes adopting a practice more widely prevalent among the more prosperous. The vital statistics of our country give no ground whatever for this belief.

Position in the U.S.A.

The American data on the connexion between reproduction and social class resemble those for Great Britain, in displaying the main lines of the phenomenon with unmistakable clarity, without supplying any means of entering into quantitative detail. The official method is to ascertain the number of children of the parents at the registration of a new birth, a method which is not at all comparable with the size of family recorded in a census. The average families, however, of mothers in the age group 35–44, should supply figures which are comparable from class to class, and, as in the case of the Scottish census, the occupational classes may be noted which have high or low averages. The following occupational lists are taken from Whitney and Huntingdon's *Builders of America*, in which work the averages have been reduced by one-third in order to obtain estimates of the average numbers of children born in completed families. This adjustment of course does not affect the order of the occupations. The year referred to by Whitney and Huntingdon is uncertain, but closely parallel results, with slightly lower fertility, are shown by the Bureau of Census report for 1925.

In the group with high fertility we have nine occupations with an estimated average family of 4·8. These are coal-miners; other miners; farm labourers; farmers; building labourers; railway labourers; railway construction foremen; metal moulders; and factory labourers, in order of decreasing fertility.

In the group with low fertility we have eleven occupations, with an estimated average family of 2·6. These, in order of increasing fertility, are, physicians; technical engineers; lawyers; book-keepers;

bankers ; teachers ; agents and canvassers ; factory officials ; insurance agents ; clerks in offices, &c. ; real-estate agents.

The group of low fertility comprises all the professional occupations except that of clergymen, who come in the next group, while the group with high fertility contains all classes of labourers, together with only three other groups, among which the rural occupation of farmers is conspicuous, in the same way as we have also seen in the French figures. On the whole the American figures show a somewhat wider range of variation than do those quoted from Scotland; for example the coal, shale and ironstone miners in Scotland show 77 per cent. more children than do the physicians and surgeons, whereas in America the miners exceed the physicians by 119 per cent. The use of data from birth registration would be expected rather to diminish than to increase the ratio, but the difference may perhaps be accounted for by two other circumstances, first that the age at marriage has been partially standardized in the Scottish figures, and secondly that the American data from birth registration refer to a much more recent date than the nearly completed families of the 1911 census in Scotland. It is not, therefore, certain that the differential incidence of the birth-rate in the United States is, absolutely, greater than that found in Great Britain, though it would be difficult to argue that it could be appreciably less.

Effects of differential fertility

The facts revealed by the vital statistics of the present century, with respect to the distribution of fertility, are in such startling opposition to the rational anticipations of the earlier sociologists, that it will not be out of place to conclude this chapter with a few reflections upon the sociological consequences, which the observed distribution of the birth-rate, whatever may be its causes, inevitably carries with it.

(i) In the first place the language of evolutionary writers has endowed the word 'success' with a biological meaning, by bestowing it upon those individuals or societies, who, by their superior capacity for survival and reproduction, are progressively replacing their competitors as living inhabitants of the earth. It is this meaning that is conveyed by the expression 'success in the struggle for existence' and this, it will probably be conceded, is the true nature of biological success. To social man, however, success in human endeavour is inseparable from the maintenance or attainment of social status;

wherever, then, the socially lower occupations are the more fertile, we must face the paradox that the biologically successful members of our society are to be found principally among its social failures, and equally that classes of persons who are prosperous and socially successful are, on the whole, the biological failures, the unfit of the struggle for existence, doomed more or less speedily, according to their social distinction, to be eradicated from the human stock. The struggle for existence is, within such societies, the inverse of the struggle for property and power. The evolutionary task to which the forces of selection are harnessed is to produce a type of man so equipped in his instincts and faculties, that he will run the least risk of attaining distinction, through qualities which are admired, or are of service to society, or even of attaining such modest prosperity and stability of useful employment as would place him in the middle ranks of self-supporting citizens. In societies so constituted, we have evidence of the absolute failure of the economic system to reconcile the practice of individual reproduction with the permanent existence of a population fit, by their mutual services, for existence in society.

(ii) The existing evidence indicates that this curious inversion exists throughout the civilized nations of the modern world. Of the eastern civilizations of India and China we are, unfortunately, not in a position to produce satisfactory statistical evidence. In the absence of European administration, the tremendous mortality among the poorest class in these countries occasioned by war, pestilence, and famine, must constitute a factor of great importance, and one that might arrest the progress of racial deterioration, though at a somewhat low level, if the conditions were otherwise similar to those observed in the West. The prolonged phase of stagnation and lack of enterprise, from which both these peoples suffered before the impact of European ideas, is at least suggestive of an equilibrium produced by these means. Among ancient civilizations the testimony of Greek and Latin writers is, moreover, unanimous and convincing, in declaring that a similar inversion was vividly evident during the period when the arts and social organization were most highly developed. We may remember that the Latin word *Proletarii* for the class of citizens without capital property meant in effect 'The beggars who have children', and that numerous, though perhaps ineffective laws, of the early Empire, were designed to encourage parentage among the higher classes.

(iii) The causes which tend to induce the inversion of the birth-rate must be deferred to the next chapter, but we may now note that these causes must have encountered and overcome great natural obstacles. Owing to the burden of parentage, and to the inheritance of wealth, its constant tendency is to make the rich richer and the poor poorer. Such a tendency is not only to be deplored in the public interest, as interfering with the distribution of wealth according to the value of the services exchanged, and will therefore be resisted in some measure by legislation, but, what is far more important, it will be resisted in the economic interests of private individuals, as would be a tendency to tax the poor more heavily than the rich. The total burden of rearing the next generation of citizens would be more easily borne, if distributed more in accordance with the ability to support it. Those who can afford most of comforts and luxuries can assuredly afford to have the most children, and upon purely economic grounds, removed from all biological considerations, the large households of the rich might be reasonably expected to produce and support more children than the small households of the poor. We have moreover no *a priori* reason to assume that mankind is an exception to the rule, which holds for other organisms, whether of plants or animals, that an increased abundance of the necessities of life is favourable to reproduction.

The indirect effect of the favourable conditions of the more prosperous classes should also be borne in mind. If the optimal level of fertility of species in general is determined by the requirements of parental care, it is a relevant fact that the superior classes have, and have had in all ages, the power to delegate a large proportion of their parental cares to others. The pressure of parental care is thereby relieved, and the optimum level of fertility raised, as in the case of the cuckoo. An obvious example is the provision of wet nurses for the care of infants, a custom which must greatly increase the prospect of an early conception by the mother. The physiological mechanism by which this is brought about may be regarded as the means by which the fertility is increased as soon as the natural burden of parental care is removed, as it would be by the death of the infant. The same principle applies to the cares required of parents in reducing the incidence of all infant and child mortality, and it is obvious that, in so far as these cares can be delegated to skilled professionals, the biological optimum of fertility, in the privileged class, will be

progressively raised. If such a privileged class were biologically isolated, its innate fertility would therefore increase to a higher level than
that found in classes who had to care for their own children, and this
apart from the fact that, for the same level of innate fertility, the
more prosperous class should, in fact, be the more prolific.

In comparatively recent times efforts have been made on a national
scale to relieve the poverty of the poor, and make a reasonable provision for the education of their children, by direct taxation of the
higher incomes and the larger estates. An important factor, if not
the whole cause, of the need for this continual redistribution of
national wealth, evidently lies in the disparity with which the burden
of parenthood is distributed and in its peculiarly heavy incidence
upon the poorest class. Enormous as is the expenditure incurred, and
important as may perhaps be its biological consequences, in relieving
the poorest class of much of the burden of parental care, and consequently, though indirectly, increasing their fertility, it would be a
mistake to suppose that the burden of parental expenditure, in any
class of self-supporting citizens, is appreciably relieved. The support
of a normal family of three or four children entails in the greater part
of our industrial population, a period of the harshest poverty, contrasting violently in comfort, security, and the possibility of saving,
with the standard of living which a childless man can enjoy on the
same wage. The economic pressure of low wages, while unable to
control the actual incidence of parenthood, with which it is in conflict, is yet able to inflict extreme hardship upon the mothers of the
next generation ; and it is a feature of our social system which should
not be overlooked, that an unduly large proportion of each generation
is derived from homes which have experienced the severity of this
conflict. The inversion of the birth-rate is a fruitful mother of social
discontent.

(iv) Intimately bound up with the concentration of reproductive
activity in the poorer classes of citizens, is a consequence, which, at
first sight, has its beneficial aspect, namely that very extensive social
promotion must take place in every generation, in order that the
numbers of the better paid classes may maintain a constant proportion
to those of the worse paid. Social promotion due to this cause must,
however, be distinguished from such normal promotion, due to increasing age and experience, as will leave the children to start with
no greater social advantages than their parents, and also from that

relative promotion of special merit or talent, which allows the more gifted, whether greatly, or in only a slight degree, a proper opportunity for the exercise of their special abilities. Promotion of these two kinds should, of course, characterize any society wisely and generously organized, whatever may be the distribution of reproduction within it; but in wishing prosperity to deserving merit we do not necessarily desire the continuous replacement of whole classes by those of humbler origin, unless such replacement can be shown to conduce to general prosperity. The sympathy which we feel, too, for the efforts of dutiful parents, to aid their children to good fortune, would be misplaced if it led us to desire the existence of increased numbers of children who require good fortune to attain a useful and prosperous way of life, and a decrease of those children who might claim such a life as their birthright.

If we take account of the natural consequences which flow from promotion throughout the whole social scale, involving the transfer of many millions in every generation, we shall find some which may be desirable, and others which are certainly harmful. In any body of people whose parents and grandparents occupied, on the average, a distinctly humbler position than themselves, it is obvious that we should not expect to find that pride of birth and ancestry, which is characteristic of many uncivilized peoples. The conservative virtue, which strives to live up to an honourable name, will be replaced by the more progressive virtue, which strives to justify new claims. At the worst pride of birth is replaced by pride of wealth, and popular sentiment tends to grant the privileges of an aristocracy, rather to the wealthy than to the well-born.

A second and less equivocal effect of wholesale social promotion is the retardation of the cultural progress of every class of the community, by reason of the need to educate up to a higher level those suffering from early disadvantages. The strain put by this factor upon our educational system seems to be severe, and it doubtless accounts, in large measure, for the slow progress of genuine culture, and for the set-backs which it seems, here and there, to be receiving. This consequence of social promotion would, of course, show itself, whatever the hereditary aptitudes of the different classes might be.

The third and most serious disadvantage is that, whereas the efficiency of the better paid classes may, in some measure, be maintained, by progressively perfecting the machinery for social promotion, in such a way as to make the best use of every grade of talent, wherever it may be found, yet the worst-paid occupations have no source from which they can recruit ability, and consequently suffer a continual degradation in the average level of the talents they possess. Without underestimating the value of the services, which those promoted perform in their new spheres, it would be fatal to ignore the extent to which the prosperity of any co-operative community rests upon the average efficiency of the great masses of citizens, and, I should add, the extent to which its well-being rests on the level at which their self-respect, enterprise and spirit can be maintained.

Summary

The different occupations of man in society are distinguished economically by the differences in the rewards which they procure. Biologically they are of importance in insensibly controlling mate selection, through the influences of prevailing opinion, mutual interest, and the opportunities for social intercourse, which they afford. Social classes thus become genetically differentiated, like local varieties of a species, though the differentiation is determined, not primarily by differences from class to class in selection, but by the agencies controlling social promotion or demotion.

Comparisons between the vital statistics, and especially between the birth-rates, of different classes, are generally defective and much out of date. The principal need in our own country is to bring the occupational classification, used in the registration of births and deaths, into harmony with that used in the census, and to record the ages of the parents in birth registration.

Numerous of investigations, in which the matter is approached from different points of view, have shown, in all civilized countries for which the data are available, that the birth-rate is much higher in the poorer than in the more prosperous classes, and that this difference has been increasing in recent generations. As more complete data have become available, it has appeared that this difference is not confined to aristocratic or highly educated families, but extends to the bottom of the social scale, in the contrast between the semi-skilled and the unskilled labourers. There is no direct evidence of a period at which the birth-rate in all classes was equal, and the decline in the birth-rate in all classes in recent decades has been apparently simultaneous, though greater in the more prosperous classes.

Since the birth-rate is the predominant factor in human survival in society, success in the struggle for existence is, in societies with an inverted birth-rate, the inverse of success in human endeavour. The type of man selected, as the ancestor of future generations, is he whose probability is least of winning admiration, or rewards, for useful services to the society to which he belongs.

If, as is still uncertain, a similar inversion has prevailed in the Asiatic centres of civilization, the mortality suffered by the poorest class must have tended to arrest the progress of racial deterioration, and had perhaps produced an equilibrium before the impact of European ideas. The condition of the Roman empire was certainly similar to that observed in modern countries.

The causes, which have produced the inversion of the birth-rate, must have been sufficiently powerful to counteract both direct and indirect economic agencies, favouring a higher birth-rate among the more prosperous. By its tendency to make the rich richer and the poor poorer, and, especially, to inflict hardship upon the parents of the next generation, the inversion of the birth-rate is an important cause of social discontent.

A consequence which, at first sight, appears beneficial, is the very large amount of social promotion which is required to maintain the proportion of the classes. Upon examination it appears that this kind of promotion should not be confused with increasing prosperity, and that it carries with it the serious disabilities of the retardation of the cultural progress of every class, and the uncompensated depletion of the poorest class in the ability to maintain their self-respect and economic independence.

XI

THE SOCIAL SELECTION OF FERTILITY

History of the theory. Infertility in all classes, irrespective of its cause, gains social promotion. Selection the predominant cause of the inverted birth-rate. The decay of ruling classes. Contrast with barbarian societies. Heroism and the higher human faculties. The place of social class in human evolution. Analogy of parasitism among ants. Summary.

May it befall that an only begotten son maintain the ancestral home, for thus wealth is increased in a house. HESIOD, eighth century B.C.

History of the theory

WE have seen in the last chapter that among civilized peoples, both in modern and ancient times, an anomalous condition has come into existence, in which the more prosperous social classes, whom we would naturally compare to the successful and well-adapted of an animal species, reproduce their kind considerably more slowly than the socially lower classes. In the ninth chapter we had seen reason to conclude that the innate and heritable disposition has in civilized man a powerful influence upon the rate of reproduction. It is now proposed to show that a logical connexion exists between these two conclusions. It is not denied that what may be called the accidents of history have from time to time determined the social position of various types of men, and have influenced the fertility of various classes. The widespread nature of the phenomenon of the differential birth-rate, existing in great bodies of people, in different nations, and reappearing after long intervals of time in entirely different civilizations, is not, however, to be explained by historical accidents. No explanation of it can be accepted, which does not flow from agencies in almost universal operation, among civilized societies of the most various types.

The theory to be here developed may be found in germ in an interesting observation noted by Francis Galton in the course of his genealogical researches; it has since been extended by several writers, step by step with the advance of our knowledge of the sociological phenomena, but neither its logical cogency, nor its importance to sociological theory, seem to have been ever widely grasped, and apart from a few ephemeral papers of my own, and a brief discussion in Major Darwin's recent book *The Need for Eugenic Reform*, it might be said to have been totally neglected.

In his book on *Hereditary Genius*, published in 1869, Galton considers the problem presented by the generally acknowledged fact that the families of great men tend, with unusual frequency, to die out. Of thirty-one peerages received by the judges of England, twelve were already extinct. Galton examined the family history of these thirty-one peerages, and lit upon an explanation which he rightly describes as 'Simple, adequate ahd novel'. A considerable proportion of the new peers and of their sons had married heiresses.

But my statistical lists showed, with unmistakable emphasis, that these marriages are peculiarly unprolific. We might have expected that an heiress, who is the sole issue of a marriage, would not be so fertile as a woman who has many brothers and sisters. Comparative infertility must be hereditary in the same way as other physical attributes and I am assured that it is so in the case of domestic animals. Consequently the issue of a peer's marriage with an heiress frequently fails and his title is brought to an end.

After giving the individual histories of these families he arrived at the following results.

(i) That out of 31 peerages there were no less than 17 in which the hereditary influence of an heiress or coheiress affected the first or second generation. That this influence was sensibly an agent producing sterility in 16 out of these 17 peerages, and the influence was sometimes shown in two, three or more cases in one peerage.

(ii) That the direct male lines in no less than 8 peerages (Galton gives the names in full) were actually extinguished through the influence of the heiresses, and that 6 others had very narrow escapes from extinction, owing to the same cause. I literally have only one case, where the race destroying influence of heiress blood was not felt.

(iii) That out of the 12 peerages that have failed in the direct male line, no less than 8 failures are accounted for by heiress marriages.

Now what of the four that remain. Lords Somers and Thurlow both died unmarried. Lord Alvanley had only two sons of whom one died unmarried. There is only his case and that of the Earl of Mansfield, out of the ten who married and whose titles have since become extinct, where the extinction may not be accounted for by heiress marriages. No one can therefore maintain, with any show of reason, that there are grounds for imputing exceptional sterility to the race of judges. The facts when carefully analysed, point very strongly in the opposite direction.

After drawing similar conclusions from other groups of peers Galton
continues:

> I tried the question from another side, by taking the marriages of
> the last peers, and comparing the numbers of the children when the
> mother was an heiress with those when she was not. I took precautions
> to exclude from the latter all cases where the mother was a coheiress,
> or the father an only son. Also since heiresses are not so very common,
> I sometimes went back two or three generations for an instance of an
> heiress marriage. In this way I took fifty cases of each. I give them
> below, having first doubled the actual results, in order to turn them into
> percentages.

TABLE 11.

Number of sons to each marriage.	Number of cases in which the mother was an heiress.	Number of cases in which the mother was not an heiress.
0	22	2
1	16	10
2	22	14
3	22	34
4	10	20
5	6	8
6	2	8
7	0	4

> I find that among the wives of peers 100 who are heiresses have 208
> sons and 206 daughters, 100 who are not heiresses have 336 sons and
> 284 daughters.

The following important paragraphs may also be quoted as showing
the weight which Galton attached to this principle.

> Every advancement in dignity is a fresh inducement to the intro-
> duction of another heiress into the family. Consequently, Dukes have
> a greater impregnation of heiress blood than Earls, and Dukedoms might
> be expected to be more frequently extinguished than Earldoms, and
> Earldoms to be more apt to go than Baronies. Experience shows this
> to be most decidedly the case. Sir Bernard Burke in his preface to the
> *Extinct Peerages* states that all the English Dukedoms created from the
> commencement of the reign of Charles II are gone, excepting 3 that are
> merged in Royalty, and that only 11 Earldoms remain out of the many
> created by the Normans, Plantagenets and Tudors.

> It is with much satisfaction that I have traced and, I hope finally
> disposed of, the cause why families are apt to become extinct in propor-
> tion to their dignity—chiefly so, on account of my desire to show that
> able races are not necessarily sterile, and secondly because it may put
> an end to the wild and ludicrous hypotheses that are frequently started
> to account for their extinction.

Alphonse de Candolle after noting these remarkable researches, and properly distinguishing between the extinction of the family and that of the male line, reasonably observes that similar conclusions must apply to the rich and affluent classes in general.

La différence de fécondité des héritières et non héritières anglaises est si grande qu'elle avertit d'une cause, jusqu'à présent inconnue, du petit nombre des naissances dans les familles aisées ou riches, de la noblesse et de la bourgeoisie. En général, les filles riches se marient aisément et selon toutes les probabilités physiologiques, confirmées par les faits que Monsieur Galton a découverts, ce sont elles qui ont la plus petite chance de laisser des descendants. Leur proportion doit donc diminuer l'augmentation de population des classes qui vivent dans l'aisance.

Unfortunately M. de Candolle associates this rational explanation of the relative infertility of the upper classes with others such as that of Herbert Spencer, which evidently belong to the category to which Galton had hoped to put an end. Nor is it quite clear that he grasped the point of Galton's argument, which is not so much that heiresses can marry more easily than other girls, but that they may more reasonably aim at a marriage which is socially advantageous, and so are liable to mingle their tendencies to sterility with the natural abilities of exceptionally able men.

In a brief but important note contributed in 1913 to the *Eugenics Review*, J. A. Cobb has given reasons for believing that the case of heiresses, observed by Galton, is but a particular instance of a far more general tendency. Restricting himself to the unconscious causes of relative infertility, Mr. Cobb points out that, just as the fortune of an heiress enables her to make a socially advantageous marriage, so among the children of parents of any class, members of the smaller families will on the average commence life at a social advantage compared to members of larger families. Alongside the many excellent qualities which enable a family to improve its social position, relative infertility also plays its part. In this way the less fertile stocks, having the social advantage, will gradually permeate the upper classes of society, and there cause the peculiar situation in which the more fortunate and successful of mankind have the smallest birth-rate.

A quotation from Mr. Cobb's paper will enable the reader to appreciate the point of view from which this important conclusion was reached.

Eugenists agree that the rising generation is largely recruited from the less fit. This is attributed partly to the fact that the upper classes marry later and partly to the fact that apart from the question of postponement of marriage the upper classes are less prolific than the lower.

There can be no doubt that at the present time the smaller fertility of the upper classes is almost entirely due to artificial limitation, but there is another cause of their smaller fertility, and it is to this that I wish to direct attention. It is important for the Eugenist to know to what cause he is to attribute this smaller fertility of the upper classes; if it is entirely due to artificial limitation, which is merely a temporary fashion, the consequences are not likely to be very serious, since the fashion for limiting the family is likely to take the usual course and spread downwards in the community, eventually equalizing the fertility in all ranks of society; or the fashion may die out altogether when its disastrous effect on the future of the race is perceived. It seems also possible that the advantage of limiting the family will appeal more to the poor than to the rich, for an additional child is a greater burden to the poor, and perhaps eventually the artificial limitation of families will have a beneficial effect on the race by reducing the size of the families of the less efficient.

If, however, as I shall try to show, there is a natural tendency under modern conditions for the more intelligent to become less fertile, the problem is a more serious one.

If variations in fertility are inherited and the wealthier classes have for generations been put through a process of selection by which members of small families have been given an advantage over members of large families, we should expect that the wealthier classes would, as a whole, be less fertile than the poorer classes.

There must be some general cause which prevents the average intelligence in a civilized community from advancing beyond a certain point. That cause seems to me to be the grading of society according to a standard of wealth. This puts in the same class the children of comparatively infertile parents and the men of ability, and their intermarriage has the result of uniting sterility and ability.

Infertility in all classes, irrespective of its cause, gains social promotion

In the development of the theory by the three writers quoted, it is evident that the progress made has consisted almost entirely in the extension of the application of the principle to a wider and wider range of social classes. This evidently followed merely from the extension of our knowledge of the classes in which an inverted birthrate manifested itself. Galton was thinking principally of titled families, because he was aware that among these the extinction of the title took place with surprising frequency; De Candolle was aware that a low birth-rate also characterized the rich bourgeois class, and perceived immediately that the principle which Galton had discovered, in the genealogies of titled families, must apply with equal force wherever social position was greatly influenced by inherited capital. It is probable that Cobb was not aware, in 1913, that the results of the census in 1911, in Great Britain, showed that the inversion was strongly developed, even in the poorest class; but he evidently has no hesitation in believing that it characterized the great mass of the population, and his note has the great merit of applying the principle to classes in which inherited wealth is relatively unimportant.

That the economic situation in all grades of modern societies is such as favours the social promotion of the less fertile is clear, from a number of familiar considerations. In the wealthiest class, the inherited property is for the most part divided among the natural heirs, and the wealth of the child is inversely proportioned to the number of the family to which he belongs. In the middle class the effect of the direct inheritance of wealth is also important; but the anxiety of the parent of a large family is increased by the expense of a first-class education, besides that of professional training, and by the need for capital in entering the professions to the best advantage. At a lower economic level social status depends less upon actually inherited capital than upon expenditure on housing, education, amusements, and dress; while the savings of the poor are depleted or exhausted, and their prospects of economic progress often crippled, by the necessity of sufficient food and clothing for their children. These obvious facts are corroborated by the arguments upon which the limitation of families is advocated; of these by far the most weighty is the parent's duty of giving to his children the best possible start in life, and the consequent necessity both of savings, and of expenditure—an argument at least as forcible for the poor as for the rich.

Selection the predominant cause of the inverted birth-rate

It has been shown in Chapter IX that not merely physiological infertility, but also the causes of low reproduction dependent from voluntary choice, such as celibacy, postponement of marriage, and birth limitation by married couples, are also strongly influenced by hereditary factors. The inclusion of the hereditary elements responsible for these, to which is certainly due the greater part of the variance in reproduction among civilized men, multiplies many-fold the efficacy of the selective principle at work. We have seen in particular that hereditary differences in the mental and moral qualities affecting reproduction, must be of such a magnitude as to produce considerable evolutionary changes, in the course of relatively short historical periods, and that such changes have in fact taken place in the moral temperament of peoples known to have been exposed to the selective action of voluntary family limitation. Consequently it is impossible to avoid the conclusion that, in a society in which members of small families are on the average at a social advantage compared to members of large families, the parents being in other respects equivalent, society will become graded, not only in respect of physiological infertility, but much more rapidly and more steeply graded in respect of those temperamental differences which conduce to celibacy, postponement of marriage and birth limitation.

This inclusive character of the theory when fully developed serves to explain a characteristic of differential reproduction, which would otherwise be quite unintelligible, namely that all the sociological factors into which differential reproduction can be analysed, alike exert their influence in the same direction. If any single sociological cause favoured contraceptive practices in the upper classes, as, for example, freer access to medical knowledge might be supposed to do, the effect to be expected would be a lower birth-rate to parents of a given age, partially compensated in the statistical aggregate by greater readiness to marry. Equally, if something inherent in the occupational conditions of the better-paid classes were supposed to induce more frequent postponement of marriage, this cause, acting alone, would be naturally accompanied by an increased birth-rate of those married in the higher age groups. What is actually observed, however, is the concurrence in the more prosperous classes, of increased celibacy, of higher age at marriage, and of a lower birth-rate of married persons of a given age; and this we should expect if the effective cause of the phenomenon were the selective promotion into higher social strata of net infertility as such, irrespective of the psychological causes which have induced it.

It is possible to test in another way whether hereditary influences supply a major and controlling cause of the differential birth-rate, or a minor and subsidiary one. Even in Galton's time many causes had been suggested for the portion of the phenomenon then known, ascribing the differences in fertility to differences in social environment, or to some assumed physiological connexion between infertility and the powers of the mind. Since then the list of suggestions has increased: they may be exemplified by, excess of food, which the upper social classes are presumably supposed to consume; excess of leisure; the stress of brainwork; the enervating influence of comfort. However baseless the supposed causes may appear to be when each is examined in detail, it is certainly possible *a priori*, that there might be some subtle influence, of the social environment of the more prosperous classes, really unfavourable to reproduction. The sharpest possible test between the two views would be to ascertain the relative fertilities, among men of a given social class, of those who had risen rapidly in the social scale as opposed to those who were born in that class. For, on the theory that we have to do principally with heritable factors affecting fertility, the fertility of the upper social classes must be prevented from rising by the lower fertility of those whom social promotion brings into their ranks; the stream of demotion of the more fertile members of the upper classes being relatively a very feeble one. Consequently, the groups enjoying rapid social promotion should, on this theory be even less fertile than the classes to which they rise.

If, on the contrary, the important causes were any of those to be included under 'social environment', we should confidently expect the families who rise in the social scale to carry with them some measure of the fertility of the classes from which they originated. We have seen that the differential birth-rate is strongly developed in the United States, and a test is therefore afforded by a table given by Huntington and Whitney of the average number of children per person in the American *Who's Who*, when the persons are subdivided according to the education they received. They are given in descending order.

TABLE 12.

Kind of education.	Estimated children per person.
College and Professional	2·4
College and Ph.D.	2·3
College	2·3
Normal, Business, Trade, Secretarial	2·3
Highschool	2·1
Elementary schools and home	2·1
Professional school only	1·9

As the total number of persons dealt with is about 25,000 the trend of these averages cannot possibly be ascribed to chance; the table appears to show unmistakably, that among Americans who attain a sufficient level of eminence to be included in *Who's Who,* those whose social promotion has been most striking have, on the average, fewer children than those whose social promotion has been less. Such a result would appear inexplicable on any of the views that connect the differential fertility of different classes with elements in their social environment, and is a striking confirmation of one of the most unexpected consequences of the theory that the dominating cause lies in the social promotion of the relatively infertile.

Whitney and Huntington give also the results of another inquiry which bear upon the same problem. They have studied the average abilities shown by Yale students coming from families of 1, 2, 3 and up to 6 or more. They find, in general, that the average ability rises as we pass from families of one, to families of two, and so on up to the largest families. This, of course, is not at all what we should expect to find, either in England or America, if we were to test the sons of the population at large. In the population generally, the classes which furnish the largest families would certainly show the lowest average scores, whether we took a scholastic, an athletic or an intelligence test. The Yale students, however, are the sons of a selected group of the population, of persons who can afford to give their children an expensive education. This basis of selection favours the more prosperous parents, who, in so far as their prosperity is due to innate causes, will be innately abler than the general population; but this selection of parental prosperity will be much more stringent for the parents of six or more children than for the parents of one only. Remembering that the only child may be expected to have a better start in life *ceteris paribus* than the member of a family of six, we may perhaps expect to find at Yale the only children of less successful parents educated at equal expense and side by side with the children of the more stringently selected parents, who are able, with families of six or more, still to send them to Yale.

It is in fact a necessary consequence of our theory of the social selection of fertility, that, whereas in the population at large fertility must become negatively correlated with such characters as intelligence, which conduce to social promotion, yet, if we could select a group of children about to enter the world with absolutely equal social opportunities we should find, within this group, fertility positively correlated with these characters. For parents who can give to a large family a certain level of educational advantage, could certainly have launched a smaller family with greater advantages, and must therefore possess a higher average level of qualities, apart from fertility, favourable to social promotion, than do the parents of smaller families whose children are actually receiving the same advantages.

I believe no investigation has aimed at obtaining a group in which the educational advantages provided by the parents are completely equalized, but the effect of a partial equalization, in raising the correlation between fertility and intelligence, is shown by some interesting data published by H. E. G. Sutherland and Godfrey H. Thomson in 1926. In an unselected group of 1924 elementary school children of the Isle of Wight, between the ages of $10\frac{1}{2}$ and $11\frac{1}{2}$ years a correlation coefficient, -0.154, with a probable error 0.023, was obtained. Since the correlation is negative and clearly significant, it appears that, with this unselected group the least intelligent children belong to the largest families. On the other hand 386 boys of the Royal Grammar School, Newcastle-on-Tyne, the intelligence quotients of which are stated to be very reliable, and 395 boys and girls of Moray House School, Edinburgh, show correlations -0.058 ± 0.04, and -0.075 ± 0.034 both of which, though still negative, are insignificant in magnitude. Since within the same school, even if of very definite class, much variation must exist in the social advantages with which different pupils enter life, these data leave little doubt that in a group in which these advantages were strictly equalized the correlation between fertility and intelligence would have been raised to a positive value. It is obvious that such results are opposed to any theory which implies a physiological opposition between fertility and intelligence, and are unintelligible upon the view that class differences of fertility are derived wholly or indeed largely from class environment.

The decay of ruling classes

The *fact* of the decline of past civilizations is the most patent in history, and since brilliant periods have frequently been inaugurated, in the great centres of civilization, by the invasion of alien rulers, it is recognized that the immediate cause of decay must be the degenera-

tion or depletion of the ruling class. Many speculative theories have
been put forward in explanation of the remarkable impermanence of
such classes. Before proceeding further it may be as well to consider
what elements of truth these suggestions may contain.

The Comte de Gobineau, to whose active mind the decline of
civilizations presented itself as the greatest of human problems,
believed that the disappearance of the original aristocracies was
universally due to race mixture. Others have not hesitated to ascribe
the infertility of the upper classes to inbreeding; but since this view
is now, on the biological side, universally discredited, it need not
be further considered. There was, on the other hand, nothing in
Gobineau's theory of the effects of race mixture which could have
been disproved by the science of his time; if the progress of genetical
knowledge has rendered his views unlikely they would still have
claims to acceptance on an historical basis, were they the only available explanation of the historical facts.

The general consequences of race mixture can be predicted with
confidence, without particular knowledge of the factors in which the
contrasted races differ. If the races and their descendants intermarry
freely, the factors which are inherited independently will be recom-
bined at random; each person of mixed race will resemble the one
parental stock in respect of some factors, and the other stock in
respect of others. Their general character will therefore be intermediate, but their variability will be greater than that of the original
races. Moreover, new combinations of virtue and ability, and of their
opposites, will appear in the mixed race, combinations which are not
necessarily heterozygous, but may be fixed as permanent racial
characters. There are thus in the mixed race great possibilities for the
action of selection. If selection is beneficent, and the better types
leave the greater number of descendants, the ultimate effect of mixture will be the production of a race, not inferior to either of those
from which it sprang, but rather superior to both, in so far as the
advantages of both can be combined. Unfavourable selection, on the
other hand, will be more rapidly disastrous to a mixed race than to
its progenitors. It should of course be remembered that all existing
races show very great variability in respect of hereditary factors, so
that selections of the intensity to which mankind is exposed would
be capable of producing rapid changes, even in the purest existing
race.

It does not seem unreasonable to conjecture that within each of the great divisions of mankind, internal adaptations or adjustments, between the different faculties of the mind and body, have been established during the long ages of selection under which these races were originally formed. If this were true, the crossing of widely different races would disturb this internal harmony and put the mixed race at an initial disadvantage, from which, even under favourable selection, it might take several generations to recover. The theory of Gobineau depends upon the effect of intermixture of a ruling aristocracy, which Gobineau invariably traces to the European stock, to whose merits he ascribes the progress and stability of all civilizations, with a mixed and coloured people inhabiting one of the great and ancient centres of population, such as China proper, or the valleys of the Ganges, the Euphrates or the Nile. Apart from his assumption of the racial origin of the rulers, the historical importance of such cases cannot be doubted; and it is in all respects probable that, as intermixture progressed, the ruling class, in the absence of selection, would entirely change its character; the qualities by which it had been distinguished would be diffused among those of the more numerous race, and in the absence of native ability, could no longer achieve the great results, for which, concentrated in a ruling class, they were originally responsible. Nevertheless, granting that the process of racial diffusion sets a certain term to any civilization which depends upon the virtues of an alien minority, it remains to be explained why great civilizations should be assumed always to rest upon so precarious a foundation. If the peoples of the world's great centres of ancient population are inherently incapable of producing of themselves a great and vigorous civilization, an explanation of so remarkable a fact is certainly required. We cannot reasonably be satisfied by accepting it as an accident, that the great civilizing races should have originally occupied only those regions of the earth, where civilization was for so long unknown.

It has been suggested that the disappearance of ruling races of foreign origin in ancient civilizations was due to the selective influence of the climate. This theory stands in the same position as that of race mixture, in respect of the main criticism to which the latter is exposed. It does not explain why the foreign element should be necessary. Apart from this it certainly appears to be a real factor in the situation. Races exposed to a new climate, and to unfamiliar diseases, are certainly in some cases at a disadvantage as regards their death-rate, and probably also as regards their birth-rate; as are the negroes in centres of population infected with tuberculosis, or Europeans in regions suffering from uncontrolled malaria.

In one respect the theory of selection by climate and disease appears to possess an advantage over that of race mixture. If the latter were the only agency at work, the disappearance of the ruling class would be accompanied by a permanent improvement of the natives. The effect of successive conquests should accumulate; so that we should expect that a people, such as the Egyptians, would be reasonably far advanced towards the type of a ruling race. The reverse appears to be the case. The effect of the selective influence of climate and disease, on the other hand, would appear to undo completely the racial benefits of an invasion. Further consideration shows that both agencies acting together would lead to an intermediate result, for in the distribution of the hereditary factors, immunity to local diseases would often be combined with the qualities of the immigrants. This consideration suggests that unless selection is directed against the qualities of the ruling race as such, some permanent improvement of the native population is a necessary consequence. On the other hand beneficial selection combined with race mixture would lead to the formation of a race combining the condition of acclimatization with the valuable qualities of the invaders. It would seem then, from the history of ancient civilizations in the East, that the rate of reproduction has never permanently favoured the ruling classes. If we consider further that, in any ordered society, the burden of maintaining the population is likely to fall upon the parents, and that the possession of wealth is likely sooner or later to be of social advantage, it is possible to offer a rational explanation, not only of the disappearance of ruling classes of foreign origin, but of the paucity of the necessary types of ability from the indigenous population. It would seem not improbable on this view that in its origin civilization was indigenous, favoured by the natural causes which admit of a dense population; that the most capable elements of this primitive civilization formed themselves sooner or later into the upper classes, that they were slowly imbued with those factors of heritable disposition which make for a reduced prolificacy; and that with the consequent development of wholesale social promotion, they were, by a rapidly increasing process, elimi-

nated from the race. Their territory, its natural resources enriched by
centuries of productive labour, and destitute of a vigorous and united
government, became the natural prey of a succession of invaders. Once
it is apparent that natural causes sufficiently explain the attenuation of
the original rulers, the disappearance of aristocracies of foreign origin
raises no new problem. The same agencies which destroyed the founders
of a civilization are capable of destroying their successors; the more
easily if the unfavourable action of selection were furthered by race
mixture and the influence of climate.

The belief that the mere existence of civilized conditions causes
degeneracy among the races which experience them, has been held
by many writers, and is strongly supported by the instances of barbarous territories which have received the civilization, and shared
the decay, of some more advanced state. The peoples of Roman Gaul
and Britain, originally formidable enough, were, during the decay of
Roman power, scarcely more capable of supporting and safeguarding
their civilization, than were the Romans; they fell a natural prey
to inferior numbers of warlike barbarians, without suffering from a
foreign climate, or from intermixture with an inferior race.

No explanation of the manner in which civilization produces this
deleterious effect, by reference to the inheritance of acquired characters, can now be regarded as admissible; and in any case it must be
doubted if the characters acquired in a state of civilization, which
include habits of industry and discipline, in addition to the training
of the intellectual faculties, can be regarded as unfitting a people to
hold its own. It is said that luxury saps the vigour, and dependence
the initiative of civilized peoples, and even if this were the unquali-
fied truth as to the effect upon the individual of civilized life, the
first generation of misfortune should restore the vigour, and give
unexampled opportunities to the initiative, of a threatened people.

We know that the tide of social promotion throughout the Roman
dominion must have flowed rapidly. The demand for men for the
Imperial services extended freely to provincials, and in the commer-
cial and industrial development of the provinces a large scope was
given to the initiative of all, even in their home towns. The upper
classes, and especially the wealthier of these, were certainly drawn
into the vortex of a cosmopolitan society, recruiting itself extensively
from the lower orders. Gauls were admitted to the Senate as early as
the reign of Claudius, who first occupied Britain. With these facts

in mind it is not surprising that after several hundred years absorption in the Roman civilization, Britain and Gaul showed scarcely greater capacity for resistance than the older provinces. For even if the conquest of Britain was a matter of difficulty to the barbarians, the very possibility of the overthrow of an organized Romano-British society shows that very inadequate use must have been made of the resources of the island.

Contrast with barbarian societies

The social selection of infertility will, it appears, characterize all states of society in which (1) distinctions of social class exist; (2) wealth is influential in determining the social position of an individual or his descendants; and (3) the economic charge of producing the next generation is borne exclusively or principally by the parents who produce them, at least in the sense that they would have been better off had they produced fewer. These conditions are of such wide application that we can have no hesitation in postulating them in all societies, ancient or modern, consisting of individuals co-operating for mutual advantage in a state of law and order. The essentially dysgenic features of the situation may, however, be expressed more aptly by saying that they are implied by the existence of a differentiation of social class in which social promotion, or demotion, is determined by the combination of two different attributes (1) socially valuable qualities, and (2) infertility. It is thus not quite exact to say that the selective agency considered must be dysgenic in any society graded according to wealth; this would imply too wide an application of the principle, for gradations of wealth in the form of the control of material property, and authority over the services of others, would seem to be inseparable from any organization of society whatever.

There have certainly existed societies, though not properly speaking civilized societies, in which the institution of social class was highly developed, in which the power and prestige of the individual rested largely upon his pedigree and kinship, in which these class distinctions were doubtless correlated with personal wealth, and in which, nevertheless, the social advantage lay with the larger families. Since in such societies we should infer from the principle of the inheritance of fertility, and the absence of any countervailing causes, that the fertility of the socially superior should be the higher, and

consequently that the powerful evolutionary force, which such difference of fertility has been shown to exert, will be directed towards
the increase of the qualities favourable to success in these societies,
and to the qualities admired in them, their importance for the study
of our theory, and for the evolutionary history of mankind, is very
great.

The state of society with which we are here concerned, which may
be exemplified by the primitive peoples of Northern Europe, as
represented in the Icelandic Sagas, in Tacitus' description of the
Germans, and probably in the Homeric poems, by the pre-Islamic
Bedouin of the Arabian desert, by many, if not all, of the Turkish and
Tartar peoples of the Central Asiatic steppes, and by the Polynesians
of New Zealand and Samoa, is characterized by a tribal organization,
influenced, or indeed dominated, by the blood feud. All these show
a strong feeling for aristocratic or class distinction, and this character,
as well as the blood feud, seems to be rather rare among uncivilized
peoples generally. For this reason it is convenient to designate this
particular type of society by a special term, which shall contrast
them with civilized, and distinguish them from the other uncivilized
peoples. We may conveniently call them barbarians. This term
is the more appropriate in that the examples given, few as they are,
include the most important groups of peoples, who have, in the
course of history, overrun the great centres of relatively permanent
civilization, and to whom the existing organization of society can be
traced back in historical continuity.

Different civilized nations, although potentially at war with one
another, may yet show essentially the same ideas, and an equivalent
development of material civilization. The cultural unity of different
barbarian tribes is usually much closer, for they are bound together
by common language, and a common oral literature, intermarry much
more frequently than can the large aggregates which constitute
civilized nations, and frequent the same fairs and festivals for trade,
religion or recreation. It is necessary to emphasize this unity of
culture because, unlike civilized societies having comparable unity,
barbarian peoples recognize private, or more properly tribal war as
a normal means for avenging and checking crime. The obligation to
avenge a kinsman was felt extremely keenly as a moral duty, to
shirk which would be incompatible with self-respect or an easy conscience, or, in Wilfred Blunt's forcible phrase as 'almost a physical

necessity'. The existence of this obligation requires that the tribes of kinsmen to which it applies shall be somewhat sharply defined, and with this obligation follows, of course, the obligation to pay, and the right to share, blood money, or to share in booty. A certain degree of economic communism thus characterizes these kindred groups, so that there is little exaggeration in saying that the economic and the military units in such societies are made to coincide. This is at least a convenient form in which to express the contrast with all civilized societies, in which the interests of the economic unit, consisting of a single individual and his dependents, may differ widely from those of the military unit, consisting of the entire nation to which he belongs. The interests of the kindred group as a whole, whose rights to life and property can only be safeguarded by military preparedness, are of course, in the first degree, founded upon military strength, and consequently, among other qualities, upon the fertility of its members.

Social position in barbarian societies consists partly in differences in kind, partly in differences in degree. A powerful group comprises not only the body of free tribesmen of authentic, or noble, pedigree, but also dependent freemen harboured by the tribe, who may have been outlawed from their own tribes, or be the remnants of tribes too weak to stand by themselves; further there may be unfree dependents, whose position only differs from that of the slaves of civilized societies, in lacking the protection of civil law, and, on the other hand, in differing but little from their masters in education or standard of living. It is important here that differences in social standing are at least as strongly felt among barbarian peoples, as in civilization, and that with them they are based not so much on occupation, as on personal and family prestige. Such differences in prestige are not, however, confined to social distinctions within the tribe, but extend to great differences in the repute and distinction of different kindred groups, according to their exploits and power.

It will be admitted, therefore, that in such barbaric societies as we have described, well-defined class distinctions are combined with a distinct social advantage of the more prolific stocks. Nor can we doubt on independent evidence that the families of the highest repute were in fact the most prolific. The high importance given to pedigree, and the care taken to preserve the names, even of a remote ancestry, is evidence of this; for such care would not generally be taken to preserve the memory of ancestors, if these were on the average less distinguished than their descendants. We can thus understand one of the factors which enables such peoples to base their social system upon blood relationship. This evidence indicates, moreover, not only a higher birth-rate but a greater natural increase, when the death-rate also is taken into account. It may well be suspected that the most eminent families suffered in war the highest death-rate, but in the severe losses which barbarian peoples suffer in times of dearth or enforced migration, the more powerful groups would certainly be at a substantial advantage; and if, as we have seen, selection would tend to increase their innate fertility above that of the less distinguished groups, it is not surprising that we should be led to conclude that society was derived generation after generation, predominantly from its more successful members.

The most important consequence of this conclusion is that human evolution, at least in certain very ancient states of society, has proceeded by an agency much more powerful than the direct selection of individuals, namely the social promotion of fertility into the superior social strata. In particular, it is important that the qualities recognized by man as socially valuable, should have been the objective of such a selective agency, for it has hitherto only been possible to ascribe their evolutionary development to the selection of whole organized groups, comparable to the hives of the social insects. The selection of whole groups is, however, a much slower process than the selection of individuals, and in view of the length of the generation in man the evolution of his higher mental faculties, and especially of the self-sacrificing element in his moral nature, would seem to require the action of group selection over an immense period. Among the higher human faculties must moreover be counted the power of aesthetic appreciation of, or emotional response to, those qualities which we regard as the highest in human nature. It will aid the reader to weigh the efficacy of the social advantage of fertility in barbarous societies, if we analyse in more detail its re-actions upon a single group of qualities typical of barbarian culture.

Heroism and the higher human faculties

The social ideas of all peoples known to us in the stage of emergence from the barbaric condition are dominated by the conception of heroism, and civilized peoples normally look back to so-called 'heroic ages' in which this conception moulded to an important degree the structure of society. The emotional influence of this idea has been so great, especially through the poetic tradition, that it is difficult to give a technically accurate characterization of the phenomenon without using terms charged with rhetorical associations. The reader must remember that we are not concerned to evaluate heroism either through praise or disparagement, but merely to consider its nature and implications as a sociological phenomenon.

The hero is one fitted constitutionally to encounter danger; he therefore exercises a certain inevitable authority in hazardous enterprises, for men will only readily follow one who gives them some hope of success. Hazardous enterprises, however, are not a necessity save for the men who, as enemies or leaders, make them so, and the high esteem in which tradition surrounds certain forms of definite imprudence cannot be ascribed to any just appreciation of the chances of success. In modern times it is obvious that a man with any immoderate heritage of this quality has an increased probability of perishing young in some possibly useful border expedition, besides an increased probability of entering an occupation not easily to be reconciled with family life. It is undeniable that current social selection is unfavourable to heroism, at least in that degree which finds it sweet as well as proper to give one's life for his country. Any great war will reveal, I believe, a great fund of latent heroism in the body of almost any people, though any great war must sensibly diminish this fund. No one will, however, doubt that in respect of prudence, long civilized peoples do differ materially from those races in which the highest personal ambition of almost every man was to win renown through heroic achievement.

An examination of the action of selection in barbarous societies in the tribal condition, reveals the possibility of the selection of the heroic qualities beyond the limits set by prudence, by a method analogous to that used in Chapter VII to explain the evolution of distasteful qualities in insect larvae. The mere fact that the prosperity of the group is at stake makes the sacrifice of individual lives occasionally advantageous, though this, I believe, is a minor consideration compared with the enormous advantage conferred by the prestige of the hero upon all his kinsmen. The material advantage of such prestige in barbarous society will, I think, scarcely be questioned; it is evident in all the heroic literature; it is directly evidenced by the deliberate vaunting of tribal achievements by professional poets; equally convincing is the great importance attached to genealogy in all such societies, by which the living boast their descent from the mighty dead. The positive aim before the hero is undying fame, he is therefore bound to all that is of good repute; to the heroic spirit all material achievements are of lesser importance. Equally important with the phenomenon of heroism itself is the esteem in which it is held.

It is inevitable in a tribal state of society that certain stocks should distinguish themselves above others in the heroic qualities. If we may assume that such qualities do in fact benefit their tribesmen, which benefit can be most readily understood through the effects of prestige, then in a tribal society heroism may become a predominant quality. In this matter sexual selection seems in man to have played a most important role. I do not here specially stress the evidence of the poetic tradition, which, in spite of the reputation of poets for effeminacy, insists on associating heroism with true love. I should rather rely on the actual marriage customs of barbarous peoples. It should be emphasized that in such marriages the political element is more in evidence than the romantic, without their being the less dominated by the emotional reactions. A marriage is likely to involve blood feud obligations; union with a powerful kindred is an essential asset. The corporate tribe is interested in the match, and sexual selection is most powerfully exerted by *tribal* opinion. The prestige of the contracting parties is all-important, and while this is partly personal, it also is largely tribal. The wooer relies upon his reputation even for the decision of the lady herself. Both in the Icelandic sagas and in the pre-Islamic poems, marriages are nearly always prompted by the political aspirations of the parties.

Such sexual selection by public opinion must influence many other qualities besides valour. Beauty, highmindedness and every other highly esteemed quality must be thereby enhanced. Its importance for us is that it influences the esteem in which the group of qualities most closely associated with heroism are held. Just as the power of discrimination of the female bird has been shown to be influenced by sexual selection *pari passu* with the ornaments which she appreciates, so in a barbarous society, in which the heroic qualities do possess an intrinsic tribal advantage, the power to appreciate and the proneness to admire such qualities will be enhanced, so long at least as reproduction is actually greatest in the predominant families. The reader who will candidly compare the current attitude towards rash actions in any long civilized society with that among the peoples under discussion, will scarcely doubt that the hero-worship of barbarous peoples was in fact a mental attitude which, however useless to modern man, played in their lives a very essential part. Changed conditions which have reversed the advantage of the heroic qualities, have also reversed the advantage of being able to recognize and appreciate them. It is obvious that the barbarous element in the tradition of our culture is that which emphasizes and indeed exaggerates, the natural inequality of man, whereas the religious and legal elements emphasize his civil equality. From the fact that the barbarians valued more highly certain qualities of human character, it is a fair inference that they perceived such differences more clearly than do civilized men. Direct evidence on this point is necessarily elusive. On questions on which we are better informed than our ancestors it is easy for us to perceive the evidences of our advantage. If the reverse were the case, it would be easier for our ancestors than for ourselves to point out the difference. The only objective fact known to me relevant to the present issue is that moderns with highly trained powers of appreciation do find in the earliest examples of extant poetry a certain elusive quality in the delineation of character, which gives to such verse a recognizable supremacy in the particular literature to which it belongs.

It is, of course, difficult to distinguish between non-percipience and indifference to the distinctions perceived. The antipodes to the spontaneous choice of emotional passion is exhibited in the account given by Risley of the importance of a University degree in the Bengalese marriage market. The successful candidate who emerges with an M.A. became instantly, and in virtue of this alien qualification, a highly eligible match, and might collect in a few months a series of endowered brides. It would be difficult to decide whether this throwing upon the board of examiners the onus of grading the candidates in *sexual* selection is due to those personal differences, which still to some extent influence European lovers, being but faintly perceived, or, although perceived, to their being esteemed of much less real importance than the University degree.

Among peoples with a considerable contribution of the barbarous element in their social tradition the predominant non-personal factor in mate selection is social class. Social class may, as we have seen, be best defined as a synthesis of such distinctions as wealth, occupation, and family as influence eligibility in marriage, taking account of these distinctions only in so far as they do in fact influence such eligibility. Historically this distinction of social class is continuous with the political and romantic prestige of the predominant clans. It is thus not an accident that the social ostentation of earlier days should survive most conspicuously in weddings. The attenuation and decay of the sentiment of class distinction is the necessary concomitant of the progressive elimination of those elements of the population which enjoy the higher rank. It will be readily understood, if the supposition is correct that among barbarian peoples the predominant class did in fact enjoy a selective advantage, that aristocratic institutions should appear to peoples recently civilized to be based on natural justice.

To summarize the points of sociological importance: (i) A barbarian people organized in kindred groups and recognizing the blood feud as the principle of social cohesion, can scarcely fail to experience a selection in favour of two qualities on which the success of the kindred group principally depends (a) the public spirited, patriotic, or heroic disposition (b) fertility. (ii) The stratification of society in these two qualities implies a selective advantage of the heroic temperament beyond the optimum advantage ascribable to prudent boldness, by reason of the social advantage of fame or heroic reputation. (iii) The power of recognizing the heroic qualities, and of conscious choice in intermarriage, introduces the dual effect of sexual selection in intensifying both the qualities selected and the communal recognition and appreciation of such qualities. (iv) This selection of the popular emotional response to the heroic qualities has the important effects of (a) stabilizing the foundations of the system by strengthening the existing basis of social cohesion, (b) intensifying the selective advantage ascribable to fame or prestige, (c) increasing the selective advantage of all qualities consciously envisaged in sexual selection, (d) exaggerating the realities of natural inequality by the development of an extreme aristocratic doctrine of hereditary nobility. It is important to notice that such practices as polygamy or servile concubinage are not in any sense primary principles of the system of causes described, but may be grafted into the system in so far as they harmonize with the prestige of the hero, or the fertility of his class. Such practices necessarily decay or are transformed to fulfil a secondary social purpose, such as domestic service, as soon as the main conditions of the system are undermined.

The place of social class in human evolution

The combination of conditions which allows of the utilization of differential fertility for the acceleration of evolutionary changes, either progressive or destructive, seems to be peculiar to man. It requires a social organism, and one which is individualistic in reproduction. Accessory factors of great importance are the elasticity of effective fertility, introduced by infanticide and other methods of family limitation, and the peculiarly heavy burden of parental care occasioned by the extraordinarily long childhood of the human race. Up to a certain stage, which was almost certainly prehuman, the ancestors of man were doubtless solitary animals, and until social life began to be developed their fertility must have been stabilized at or near the optimum appropriate to the requirements of parental care. It is to be presumed that infanticide came to be practised, in consequence of an increasing foresight of impending hardships, at an exceedingly remote period, perhaps early in man's history as a social animal. During the immense period of early social life he appears to have learnt to co-operate with his fellows, probably by sympathy with their expressed emotions, instinctively to shun social opprobrium, and perhaps to improve his chances of posterity by mate selection. The sentiment of preference, seems at least so essential to the sexual instincts of man that it is difficult to doubt that sexual selection was early established in mankind, though it may be that the special conditions experienced by their barbarian ancestors prejudice the opinion of civilized man in this matter. Apart from this factor, the evolution of man in the early social stages must have been directed principally towards the establishment of the characteristics favouring individual survival in the social environment.

At some stage, the period of which it would be useless to conjecture, societies must have come into existence, in which some degree of continuity of social class was ensured by the inheritance of property, privilege, prestige, or social function, and in which individual differences in the socially valuable qualities were strongly appreciated, and allowed to contribute towards the assessment of social rank. If these characteristics, perhaps in quite a rudimentary form, were combined with a tribal organization of society, in which the rights of each kindred group were established ultimately upon its military power, so that the most fertile were mingled with the most admired or eminent strains in the predominant clans, a profound change must have come over the speed and direction of evolutionary progress in such societies. The new force acted through, and by means of, man's appreciation of human excellence in his fellow men. This appreciation, both in the social selection, and in the sexual selection which formed a part of it, was doubtless guided, from the first, by the interests of the tribal group, or of society as a whole. Such a process provides the means of relatively rapid evolutionary progress in qualities which subordinate the individual interest to that of others, the evolution of which, merely by the selective elimination of entire societies, would seem to be extraordinarily slow. It has, however, all the dangers noted in the case of sexual selection, of running into extravagance; for the standards of taste will necessarily be modified, step by step, with the qualities to which preference is given, and it may, indeed, be that some of the virtues which appeal most to our imaginations are more of the nature of ornaments than of serviceable moral ideals. Among the qualities which seem to have been extravagantly developed among all barbarian peoples are personal and family pride, and, but for the sharp lessons which must always be learnt in a harsh environment, made harsher by mutual warfare, extravagances of this kind might, it would seem, have become much more common than they are in man's moral nature.

It would be of great interest, if it were possible, to compare the rapidity of progress among barbarian peoples, with that of the decline among the civilized. The variations available for evolutionary changes of a destructive nature must be so much more abundant than those available for progressive changes, that we might expect selective intensities of the same order to take some hundreds, or perhaps a thousand generations, to build up in their perfection attributes of the mind, which ten generations of adverse selection might demolish. It does not appear that there should be any great difference in the selective intensities developed in the two cases, if each were in a steady state. The effect of differences of fertility would be increased in the barbarian condition, and diminished among civilized man, by the direct action of prosperity in favouring reproduction, and this might seem to make the favourable selection among barbarian peoples more intense than the unfavourable selection among the civilized. Against this should, I believe, be set the fact that, although uncivilized peoples, by practising infanticide, can produce as great a variation in net fertility as can birth limitation in civilized man, yet the selective effect of these differences in fertility must, in a harsh environment, be much diminished by heavier mortality in infancy and childhood, so that the net differences of fertility available may be less. With respect to the action of sexual selection it might be thought that this factor, while accelerating progress among barbarians, must be retarding decay among the civilized. But this is certainly not the case in a society in which the prospects of fertility diminish with social advancement, for in such a society it will, on the average, be biologically advantageous, to make, of two possible marriages, that which is socially the less eligible. Moreover those who are most particular in the choice of a mate will most frequently diminish their fertility by postponement of marriage or celibacy. Consequently sexual selection must be judged to intensify the speed of whichever process, constructive or degenerative, is in action. The intensity of this influence must, however, be much diminished in the later stages of civilized societies, with the decay of the appreciation of personal differences.

Analogy of parasitism among ants

The reaction of economic causes upon the distribution of fertility in human civilization is so disastrous that we could scarcely expect to find it adequately paralleled in insect societies, for among these an active intercommunal selection appears always to be possible. Nevertheless the interpretation put by Forel upon the reaction of the ant *Tetramorium cespitum* to a rather rare parasitic ant *Strongylognathus testaceus* is sufficiently apposite to be compared to the human situation. The structure of the parasitic ant suggests that it was formerly a slavemaker, but it is now too feeble to be an effective combatant, and the mixed colonies rely for defence upon the host workers. The parasite neuters, though able to excavate and to feed independently, contribute little or nothing to the structure of the nest, and probably obtain most of their food from the tongues of the hosts. They take no part in the care of the young, even of their own queen, and, being thus apparently a survival useless to their own species, it is not surprising that they are produced in relatively scanty numbers compared to the males and queens, which are very small and produced in abundance. The sexual forms of the host, on the other hand, are relatively enormous, since the queens have to supply the biological capital for founding new and independent colonies. In most cases of ant parasitism the mother of the parasitized community is in some manner or other eliminated, and the parasitism consists in the exploitation of the social instincts of the surviving workers for rearing an alien brood. The case under consideration is peculiar in that it appears to be well established that the host queen continues to live and, in addition to the parasite queen, to lay, in the parasitized colony; but that in this condition she never produces fertile females, the parasites thus gaining a continual supply of host workers to house, feed, educate and defend them, while the fertile queens issuing from the nest are of the parasite species only. Forel ascribes this remarkable condition to the regulatory or economic instincts of the host workers, for the females and males of the parasite are smaller and less troublesome to nourish. This, he says, is evidently sufficient to induce the host workers to rear them in place of their own enormous queens and males, the larvae of which they therefore undoubtedly devour or neglect, as they do in the case of all that seems to be superfluous.

Whether or not Forel is right in this interpretation, his suggestion illustrates well the effect of economic law, if allowed blindly to act in the regulation of fertility. Since the parasites are found in only a small proportion of the *Tetramorium* nests, these insects have presumably some defensive instincts, which usually succeed in resisting infestation. In human societies man is his own parasite, a circumstance which seems to ensure that all civilized societies shall be fully infested.

Summary

In accordance with the theory, developed with successive extensions by Galton, De Candolle, and J. A. Cobb, it is shown that the inversion of the birth-rate is a consequence of two causes which have now been fully demonstrated: (i) The inheritance of the characters, whether physical or psychological, determining reproduction; (ii) The social promotion of the less fertile. The various theories which have sought to discover in wealth a cause of infertility, have missed the point that infertility is an important cause of wealth.

In the light of this theory we can understand how it is that the more prosperous classes, not only have fewer children when married and at a given age, but that they also tend to marry later in life, and to remain more frequently celibate, than do less prosperous classes. We can understand how it is, although the poorer classes generally have more children than the rich, yet that persons of distinction, who have enjoyed great social promotion, are found to have fewer children than persons of equal distinction who have been less promoted. Again, although among the general population the larger families of the less successful classes produce a negative correlation between the ability of a child, measured in various ways, and the size of the family to which he belongs, yet in selected groups of children, chosen as receiving more equal social opportunities, we should expect to find, as indeed is found, with the Yale students, and to a less degree with some middle-class English schools, that the correlations tend to be raised to a positive value. All these facts would be highly paradoxical upon the view that the differences in fertility were the direct result of differences in social environment.

In the problem of the decay of ruling classes it is shown that neither race-mixture, nor the selective action of climate and disease, would suffice to explain their failure under favourable selection. The causes to which we have traced the inversion of fertility must have been operative in the most ancient civilizations, as in our own, and serve to explain the historical importance of ruling races, through the absence of the proper attributes in the native populations. The same causes ensure an adverse selection acting upon each conquering people in turn.

The decline of barbarian peoples, which have received the civilization, and shared the decay, of some more advanced state, without suffering from a foreign climate or from intermixture with an inferior race, is intelligible by the social promotion and extinction of their more capable members.

Certain uncivilized peoples characterized by a tribal organization, the blood feud, and the importance attached to kinship and pedigree, exhibit a state of society in which the more eminent are certainly the more fertile, and in which the effects of Natural Selection are greatly enhanced by social and sexual selection. The action of these factors is of particular importance in respect of the qualities recognized by man as socially valuable, which have, in this way, received a selective advantage very much greater than any which could be ascribed to the differential elimination of entire tribes. The group of qualities understood by these barbarian peoples as associated with heroism has thus been developed considerably beyond the optimum of individual advantage. The higher mental qualities of man, and especially his appreciation of them, seem to be ascribable to the social selection of this type of society.

The selection in favour of the qualities admired among barbarian peoples was probably almost as intense as that in the opposite direction among the civilized. The destructive effect of the adverse selection on these qualities, must, however, be much more rapid than the process by which they were built up.

If the opinion of Forel be accepted as to the reaction of the host ant *Tetramorium Cespitum* to its parasite *Strongylognathus testaceus* the situation established in parasitized colonies bears some analogy to the economic reactions of civilized man towards reproduction. Whereas, however, the majority of *Tetramorium* nests keep themselves free from parasites, human societies inevitably show the necessary variations to ensure their own infestation.

XII

CONDITIONS OF PERMANENT CIVILIZATION

Apology. A permanent civilization not necessarily unprogressive. Redistribution of births. Social promotion of fertility. Inadequacy of French system. Problem of existing populations. Summary.

And nothing 'gainst time's scythe can make defence
Save breed to brave him when he takes thee hence. SHAKESPEARE.

But a good will is as a mighty god. SOPHOCLES.

Apology

IT is to be supposed that all scientific men accept, at least in theory, the view that the advancement of biological knowledge will ultimately be of service in the practical affairs of mankind; but there can be little doubt that most would feel some embarrassment if their views had to be made the basis of practical, and therefore controversial, policy. There is good reason for this; for scientific men, no less than others, imbibe, before their profession is chosen, and perhaps later, the same passions and prejudices respecting human affairs as do the rest of mankind; and while the standard of critical impartiality necessary to science may, perhaps, be easily maintained, so long as we are discussing the embryology of a sea-urchin, or the structure of a stellar atmosphere, it cannot be relied on with confidence in subjects in which these prejudices are aroused. Any failure in this respect is, moreover, harmful to Science. The enthusiast whose sole interest in science is to find support for a preconceived social policy, not only deceives himself, and perhaps others, who are not aware of his bias, but inevitably brings some discredit upon all other workers in the same field. The investigator undoubtedly best preserves both his scientific reputation and his peace of mind, who refrains from any opinion respecting human affairs, and declines the responsibility for any practical action which might be based upon the facts he has brought to light. If such a one were to admit that mankind might, with advantage, make more use of scientific knowledge of all kinds, his practical policy would, apparently, go no further than to advocate a more general diffusion of this scientific knowledge, its application to mundane affairs being relegated to others (unspecified) who might make it their particular business.

275

The reader will have realized that our conclusions as to the selective agencies at work in civilized societies come perilously near to being capable of practical application. Conscious as I am of the inconveniences of such a course, it would appear to me merely cynical if, having established, as I believe, the main cause of the instability of human civilizations, I were not to attempt at least to specify the conditions of permanence for a civilization similar to our own. If these conditions should seem to involve far-reaching changes, difficult to be brought about, I trust this will not be imputed to the very academic irresponsibility which I am anxious to deprecate, but to the deep-seated nature of the cause to which the instability has been traced.

A permanent civilization not necessarily unprogressive

By a permanent civilization it is evident that we should not mean a rigid or a stagnant one. Our current civilization does already, no doubt, embrace the entire earth, although it has not yet experienced the final reactions to it of several important bodies of people. We may, however, grant that it has no competition to fear, without admitting that, to be permanent, it need not be progressive. For, so long as it exists, progress in the sciences and arts will certainly change the environment in which men live, and thereby change the social value of different kinds of men. It is, in fact, a necessary condition of the permanence of a civilization resembling our own, that the human race should be capable of making biological progress in whatever direction may be required by the current state of society. Even if it were argued, therefore, that a rigid system of occupational castes, each compelled to bear the burden of its own necessary reproduction, would ensure biological permanence, much as the permanence of a society of genes is ensured in cell-division, yet this could not be admitted as solving the problem of a permanent civilization. On the other hand, this phrase need not be taken to imply an absolute immunity against destruction from all causes, but only such mundane immortality as is enjoyed by one-celled animals and plants, which may perish, and indeed do in great numbers, but which have, in their inner constitution, no inevitable causes of senility and decay.

Redistribution of births

The most obvious requirement for a society capable of making evolutionary progress, in accordance with its current needs, is that reproduction should be somewhat more active among its more successful, than among its less successful members. In comparison with the actual distribution of reproduction found in civilized societies, such a condition, apart from the reason for which it is propounded, or the means by which it may be brought about, would seem to be attended by considerable advantages. Extreme poverty would certainly be mitigated, and a more even distribution of wealth insensibly brought about. There would be less need for the process of transferring wealth from the rich to the poor by direct taxation, and this should be welcomed by those who believe that this process is in practice extremely inefficient, in that the real wealth transferred to the poor is much less than the loss incurred, directly and indirectly, by the rich. There is, I believe, only one undesirable consequence which would seem, at first sight, to follow from a redistribution of the incidence of reproduction, namely that the accumulation of capital, now possible in the wealthier classes, might be diminished. Without denying that this might be an effect of a more adequate birth-rate among the wealthier classes, it is possible at least to see a countervailing advantage; for if there were less free money for investment, or, at least, if a rise in interest rates were necessary to induce the same rate of accumulation of savings, the capital withdrawn from the investment as money, would, in fact, have been invested in the production of men of average ability somewhat above that of the general population, and adequately educated for their work in life.   This is a form of capital which also earns dividends, and is, indeed, necessary for the profitable investment of all other capital.   It may be suspected that wealth would accumulate just as rapidly as at present, only more in the form of capable men, and perhaps less in bricks and mortar designed for hospitals and reformatories.

Social promotion of fertility

A moderate superiority of upper class fertility would, if we accept the argument of Chapter XI, be established, without interference with personal liberty, by a moderate tendency towards the social promotion of the more fertile strains in the population, using this phrase to designate those who, whether physically or psychologically, are more inclined to bring children into the world. Different as such a condition seems from the incidence of the system of wages and salaries current in this country, it certainly appears on examination that our wage system could with some advantage be modified in this direction, both from the point of view of the purely economic function of the wage system, and from that of current canons of natural justice. If, postponing the consideration of the machinery by which equitable arrangements are made between different employers, we contrast, with the utmost simplicity, the two systems by which the amounts received by the employed may be determined, (A) equal pay for equal work, implying considerable differences in the standards of living, of industrially equivalent persons, supporting families of different sizes, or, measured in the realities of consumption and savings, a bonus for childlessness of about 12 per cent. of the basic wage, for each child avoided, against (B) an equal standard of living for equal work, by the inclusion in the wage of a family allowance equivalent to the actual average cost of the children, it will be seen that the second system (B) while abolishing the social promotion of effective sterility, possesses the economic advantage of increasing the inducement to increased efficiency. The wage system is simplified for the recipient, whose standard of living now depends solely upon the value of the services which he can perform; regarding the system of differential wages or salaries as a means by which the community, either directly or through the private employer, induces the individual to perform socially valuable services, its capacity is now wholly exerted towards this end, instead of being, as in the first system (A) partially dissipated in inducing him also to refrain from parenthood.

A principal motive for the adoption of the systems of family allowances, now almost universal in France, was the need of making provision for the children of the poorer wage-earners in the period following the war of 1914–18, when it was felt that French industry could not bear the burden of maintaining a satisfactory standard of living among the workers, if it were also saddled with the necessity of supporting a higher standard of living for those among them who were childless.

In this respect the French system may be regarded as aimed at relieving poverty without increasing the burden of wages borne by the industry; though it is probable that, in the outcome, the saving effected by the redistribution of wages has been divided between the work-people, as wage earners, and the industries which they serve, the advantage to these being in part handed on to the general body of consumers in the form of lower prices. As soon as it became realized that these advantages could be gained, by the simple

expedient of establishing equalization pools among the different employers, it is easily understood that the system should have spread
rapidly. The actual amounts paid into or out of these pools on account
of individual employers are of course small, for any large employer
will support out of wages a number of children nearly proportional
to his total wage bill; they play, nevertheless an essential part in the
working of the system, by removing from the individual employer
the economic inducement, which he would otherwise have, to give
employment preferentially to childless persons. Any employer who
happens to be supporting, through the family allowances included in
his wage bill, more than the average number of children supported
by his association, will, therefore, recoup himself, by means of the
equalization pool, from other employers who happen to be supporting
less than the average number of children.

A second motive, which was important in establishing a family
allowance system in France, was the national desire to check the
tendency towards a diminution of population through the insuffi-
ciency of the number of births to replace the existing adults. The
decline of the birth-rate had begun somewhat earlier in France than
elsewhere, and its onset had been more gradual, so that it was more
widely realized among the French people than it is now in other
European countries, and especially in Great Britain, which, after
a more sudden fall, are now failing much more considerably than is
France to maintain the numbers of their existing populations. Whereas, however, the economic objects of the French system, in combating
unmerited poverty, in inducing industrial contentment, and in giving
full employment to the industrial population, seem to have been
very satisfactorily realized, the biological object of increasing the
number of births has met, so far, with no appreciable success, even
in those associations in which the allowances for the later children
are sufficient to give a positive economic inducement to further
parentage. When account is taken of the age distribution of the adult
population, the supply of children fell short, in 1927, by about 9 per
cent. of the number needed to maintain a stationary population, and
this shortage has been increasing in recent years, although much
more slowly than in England, Germany or Scandinavia. A recent
estimate for Western and Northern Europe as a whole, in 1926, shows
a deficiency of 7 per cent., and in view of the rapidly falling birth-
rate among the larger populations in this area, it is certain that this

deficiency is increasing rapidly. In 1927 it appears to have been about 18 per cent. for England and Wales.

It is arguable, therefore, but by no means certain, that the incorporation of family allowances into the French wage-system has exerted some influence towards checking the tendency to a decrease in population; though, if it is ultimately to prove itself effective in this respect, it must be admitted that its action is remarkably slow. Whether it will prove more effective to a generation that has grown up under its influence, and whether it is largely inoperative owing to the example of better paid classes, to whom the system has so far been very inadequately applied, further experience can alone determine. Its importance for us is that it provides the only known means, which has a rational expectation of proving effective, of controlling a tendency to population decrease, which has already gone alarmingly far in Northern and Western Europe, and may be foreseen with some confidence for English-speaking peoples in other parts of the world.

We need not, of course, here consider the problem of the regulation of population density in general, but need only notice that the objection that the wage system must be used to encourage childlessness, because no other means are sufficiently powerful to restrain over-population, is not one that can weigh with those familiar with modern tendencies.

Inadequacy of the French System

For the purpose of counteracting the social advantage of the less fertile, a purpose which was, of course, far from the minds of its founders, the system of family allowances now established in France, must be judged to be very inadequate. This is particularly true of its failure to establish family allowances proportionate to the basic wage throughout the salaried occupations, for, in the absence of such a provision, the tendency to eliminate the higher levels of intellectual ability, which show themselves particularly in these occupations, must be almost unchecked. For a permanent and progressive civilization we must postulate that this defect should be remedied, a requirement which is the more moderate since the system presents the same economic advantages to whatever income levels it be applied. A more serious difficulty seems to be presented in counteracting the social promotion of the less fertile in occupations not in receipt of fixed wages or salaries, but rewarded by fees or payments from many different individuals. Save in a population anxious to do the best for itself, both biologically and economically, the economic situation of these occupations must present a difficulty; though with goodwill, and in pursuit of an intentional policy, the difficulty would seem to lie rather with the establishment, than with the working, of a system of mutual insurance, which would provide a redistribution of income equivalent to that experienced by the earners of wages and salaries. If the bulk of the population were convinced by experience of the advantages of the system to those involved, it would be surprising if the remainder should prove themselves unable to devise and adopt co-operative schemes suitable to their own needs.

The foregoing paragraphs will have served their purpose if they have put before the reader, in a sufficiently definite and concrete form, the proposition that it is not inherently impossible for a civilization essentially similar to our own to be so organized as to obviate the disastrous biological consequences to which our own seems to be exposed. It is to be presumed that much more advantageous proposals could be elaborated. My point is merely that the biological difficulty, though intimately related to economic organization, is not imposed upon mankind by economic necessity.

Problem of existing populations

If we turn from the relatively easy problem of specifying the conditions, on which a human society might avoid the unfortunate biological consequences of the selective agencies, by which human societies are threatened, to the problem of establishing an equally advantageous condition, among the populations of our existing civilization, it is obvious that we shall be faced with the necessity for far more drastic methods than those which would suffice for the easier problem. For existing populations are already stratified in respect of the various innate characters which have, in recent centuries, favoured social promotion. This has been demonstrated, in the case of intelligence, by the comparison of the average scores attained, by the children of different occupational groups, in the tests judged by psychologists to be most suitable for gauging intelligence. What is, perhaps, more important, is that a number of qualities of the moral character, such as the desire to do well, fortitude and persistence in overcoming difficulties, the just manliness of a good leader, enterprise and imagination, qualities which seem essential for the progress, and even for the stable organization of society, must, at least equally with intelligence, have led to social promotion, and have suffered, in consequence, a relative concentration in the more prosperous strata of existing populations. These classes have, however, also been selected for relative infertility. We have seen reason to believe that they must be, to some extent, congenitally averse to the consequences of normal reproduction in existing economic conditions, and to some extent, also, to actions such as early and relatively imprudent marriage, which are favourable to reproduction. They are now producing, in countries such as England, probably less than half the children needed to maintain their numbers, and there can be no doubt that this fraction is still decreasing somewhat rapidly. The reformer must expect to encounter deep-seated opposition in the classes on which he would naturally rely for an intelligent anxiety for the future of their country, owing to the fact that many in these classes owe the social promotion of their forbears, and their present prosperity, less to the value of their services to society than to a congenital deficiency in their reproductive instincts. While it is certain, however, that opposition will be experienced from this source, it cannot be affirmed with certainty that members of the more prosperous classes in receipt of salaries would not respond to an adequate system of family allowances, sufficiently to maintain their numbers, after an initial loss of a considerable fraction of the valuable qualities now concentrated in these classes. The decline in fertility has been more rapid among the more prosperous classes than in the general population and, in so far as this is due to these classes being more sensitive to economic and prudential considerations, we might reasonably anticipate that their response to family allowances on a scale adequate to meet the actual expenditure incurred in respect of children, might be more rapid and ultimately greater than that of the other classes. If this were so, and if, at the same time, preferential promotion of infertile strains from the less prosperous classes were entirely to cease, it seems not impossible that the fertility of the upper classes might be restored, by the differential elimination of the less fertile strains, within no very lengthy historical period.

The length of the time necessarily required before the present agencies causing racial deterioration could be completely annulled, presents, perhaps, the most formidable obstacle to such an attempt. It would be contrary to all experience to suppose that the history of the next two centuries will not be broken by violent political vicissitudes. The policy of every nation will doubtless, from time to time, fall under the sway of irrational influences, and the steadfastness of intention necessary to maintain a policy in pursuit of a remote, though great, advantage, however clearly the reasons 'for that policy were perceived, is not to be expected of every people. Peoples naturally factious and riven by mutual distrust cannot well exert, or even attempt, any great effort. This, however, is no reason for men of goodwill to abandon in despair such influence for good as is placed within their reach.

A people among whom no rational motive for reproduction is widely recognized, but who are principally urged thereto by imprudence and superstition, will, inevitably and rapidly, become irrational and fanatical. The experience of previous civilizations should warn us of the probability of the increase of these elements among modern peoples. We should not ignore the possibility that even a rightly directed policy, adopted with enthusiasm, might be frustrated by this cause before it had time to work its effects. This danger would perhaps be minimized if the knowledge that parenthood, by worthy citizens, constituted an important public service, were widely instilled in the education of all, and if this service were adequately recognized as such in the economic system.

A danger of a more intellectual kind lies in the growing tendency towards a divorce between theory and practice. Specialization in the sphere of intellectual endeavour is necessary, and justified by the limitations imposed by nature upon our individual abilities. There is less justification for the thinker to detach himself from the natural outcome, in the real world, of his theoretical researches. Such detachment sterilizes theory as much as it blinds practice. It carries with it the natural, but unfortunate consequence, that the men of will and energy whose business is practical achievement, having learnt a natural contempt for irresponsible theorists, should also ignore the practical guidance afforded by theoretical considerations.

It would be idle to speculate upon the probability that our civilization should attain that permanence which has been denied to all its predecessors. The thought that the odds may be heavily against us should not influence our actions. The crew of a threatened ship are not interested in insurance rates, but in the practical dangers of navigation. Without ignoring these dangers we may perhaps gain some hope in that they are not wholly uncharted.

Summary

A redistribution of births, apart from the reason for which it is propounded, or the means by which it may be brought about, would be attended by economic advantages.

A moderate social promotion of fertility, such as should maintain a favourable birth-rate, is not incompatible with the economic organization of our own civilization, and would provide a means of combating the current tendency to a decrease of population.

The system of family allowances adopted in France is, however, definitely inadequate to preserve the higher levels of intellectual ability. There is, at present, no reason to doubt that a permanent civilization might be established on a more complete system.

The composition of existing populations, graded both in social ability and in effective infertility, presents special, and much graver difficulties, which only a people capable of deliberate and intentional policy could hope to overcome.

WORKS CITED

H. W. Bates (1862). 'Contributions to the Insect Fauna of the Amazon Valley.' *Trans. Linn. Soc. Lond.* xxiii, p. 495.

W. Bateson (1894). *Materials for the Study of Variation*. London.

W. Bateson (1909). 'Heredity and Variation in Modern Lights'. *Darwin and Modern Science*, pp. 85–101. Cambridge: University Press.

W. S. Blunt (1903). *The Seven Golden Odes of Pagan Arabia*. Chiswick Press.

A. de Candolle (1873). *Histoire des Sciences et des savants depuis deux siècles*, pp. 157–61. Geneva.

P. Cameron (1882). *Monograph on the British Phytophagous Hymenoptera*, vol. i. London: Ray Society.

W. E. Castle (1914). 'Piebald Rats and Selection; an Experimental test of the Effectiveness of Selection and of the Theory of Gametic Purity in Mendelian Crosses.' *Carnegie Inst. Publ.* No. 195, 56 pp.

J. A. Cobb (1913). 'Human Fertility.' *Eugenics Review*, iv, pp. 379–82.

C. Darwin (1859). *The Origin of Species*. London: John Murray.

C. Darwin (1868). *Variation of Animals and Plants under Domestication*. (Chapter xxii and p. 301.) London: John Murray.

C. Darwin (1871). *The Descent of Man, and Selection in Relation to Sex*, p. 399. London: John Murray.

F. Darwin (1887). *Life and Letters of Charles Darwin*. London: John Murray.

F. Darwin (1903). *More Letters of Charles Darwin*. London: John Murray.

F. Darwin (1909). *Foundations of the 'Origin of Species'*, pp. 2, 77–8, 84–6.

L. Darwin (1926). *The Need for Eugenic Reform*. London: John Murray. 529 pp.

F. A. Dixey (1908). 'On Müllerian Mimicry and Diaposematism.' *Trans. Ent. Soc.*, 1908, pp. 559–83.

J. C. Dunlop (1914). 'Fertility of Marriage in Scotland.' *Journ. Roy. Stat. Soc.* lxxvii. 259.

A. S. Eddington (1927). *The Nature of the Physical World*.

R. A. Fisher (1922). 'On the Dominance Ratio.' *Proc. Roy. Soc. Edin.* xlii, pp. 321–41.

R. A. Fisher and E. B. Ford (1928). 'The Variability of Species in the Lepidoptera, with reference to Abundance and Sex.' *Trans. Ent. Soc.*, 1928, pp. 367–84.

E. Ford and H. O. Bull (1926). 'Abnormal Vertebrae in Herrings.' *Jour. Marine Biol. Ass.* xiv, pp. 509–17.

Forel (1900). *Mittl. Schweiz. Ent. Gesell*, x, pp. 267–87.

J. G. Fraser (1909). *Psyche's Task*. London: Macmillan & Co.

J. C. F. Fryer (1913). 'An Investigation by Pedigree Breeding into the Polymorphism of *Papilio polytes*.' *Phil Trans.* B, vol. cciv, pp. 227–54.

F. Galton (1869). *Hereditary Genius*, pp. 123–33.

J. H. Gerould (1923). 'Inheritance of White Wing Colour, a Sex-limited (Sex-controlled) Variation in Yellow Pierid Butterflies.' *Genetics*, viii, pp. 495–551.

E. Gibbon (1776). *The Decline and Fall of the Roman Empire*, chapter xiv.

Le Comte de Gobineau (1854). *Essai sur l'inégalité des races humaines*. Paris: Firmin-Didot et Cie.

D. Heron (1914). 'Note on Reproductive Selection.' *Biometrika*, x, pp. 419–20.

D. Heron (1906). *On the Relation of Fertility in Man to Social Status*. London, Dulau & Co.

F. L. HOFFMAN (1912). 'Maternity Statistics of the State of Rhode Island, State Census of 1905.' *Problems in Eugenics.* London: Chas. Knight.

H. E. HOWARD (1920). *Territory in Bird Life.*

E. HUNTINGTON and L. F. WHITNEY (1928). *The Builders of America.* London: Chapman & Hall.

T. H. HUXLEY (1854). 'On the Educational Value of the Natural History Sciences.' *Collected Essays,* vol. iii, p. 46.

D. F. JONES (1928). *Selective Fertilisation.* University of Chicago.

G. H. KNIBBS (1917). *Census (1911) of the Commonwealth of Australia.* Melbourne: McCarron, Bird & Co. 466 pp.

D. E. LANCEFIELD (1918). 'An Autosomal Bristle Modifier Affecting a Sex-linked Character.' *Amer. Nat.* lii, pp. 462–4.

RAY LANKESTER (1925). 'The Blindness of Cave Animals.' *Nature,* cxvi. 745.

L. MARCH (1912). 'The Fertility of Marriages according to Profession and Social Class.' *Problems in Eugenics,* p. 208. London: Chas. Knight.

G. A. K. MARSHALL (1908). 'On Diaposematism, with Reference to some Limitations of the Müllerian Hypothesis of Mimicry.' *Trans. Ent. Soc.* (1908), pp. 93–142.

G. J. MENDEL (1865). 'Versuche über Pflanzen-Hybriden.' *Verh. Naturf. Ver. in Brünn,* vol. x.

T. H. Morgan, C. B. Bridges, and A. H. Sturtevant (1925). 'The Genetics of Drosophila.' *Bibliographia Genetica,* vol. ii, pp. 1–262.

F. Müller (1879). 'Ituna and Thyridia: a Remarkable Case of Mimicry in Butterflies.' *Trans. Ent. Soc.,* 1879, p. xx.

K. PEARSON (1899). 'Mathematical Studies in Evolution; VI Genetic (Reproductive) Selection.' *Phil. Trans.* A, vol. cxcii. 257.

PLATO. *The Republic,* Book V.

PLINY. *Natural History,* Book 29, chapter iv.

POLYBIUS. *Histories,* Book XXXVII.

E. B. POULTON (1908). *Essays on Evolution.* Oxford: Clarendon Press.

E. B. POULTON (1924). 'Mimicry in the Butterflies of Fiji considered in Relation to the Euploeine and Danaine Invasions of Polynesia and to the Female Forms of *Hypolimnas bolina* L., in the Pacific.' *Trans. Ent. Soc.* (1924), pp. 564–691.

A. O. POWYS (1905). 'Data for the Problem of Evolution in Man. On Fertility, Duration of Life and Reproductive Selection.' *Biometrika,* iv, p. 233.

R. C. PUNNETT (1915). *Mimicry in Butterflies.* Cambridge: University Press.

C. TATE REGAN (1925). *Organic Evolution.* Presidential Address to Section D of the British Association, 1925.

H. H. RISLEY (1891). *Tribes and Castes of Bengal.*

G. C. ROBSON (1928). *The Species Problem.* Edinburgh: Oliver & Boyd.

J. SCHMIDT (1919). 'La valeur de l'individu à titre de générateur, appréciée suivant la méthode du croisement diallèle.' *Comptes Rendus des travaux du laboratoire Carlsberg,* vol. xiv, No. 6, pp. 1–33.

SENECA. *Of Consolation to Helvia,* 16.

W. ROBERTSON SMITH (1885). *Kinship and Marriage in Early Arabia,* p. 157.

T. H. C. STEVENSON (1920). 'Fertility of various classes in England and Wales from the Middle of the 19th Century to 1911. *Jour. Roy. Stat. Soc.,* vol. lxxxiii, p. 401.

H. E. G. SUTHERLAND and G. H. THOMSON (1926). 'The Correlation between Intelligence and Size of Family.' *Brit. Jour. of Psychology,* xvii, pp. 81–92.

C. F. M. SWYNNERTON (1915). 'A Brief Preliminary Statement of a few of the Results of Five Years' Special Testing of the Theories of Mimicry.' *Proc. Ent. Soc.* (1915), pp. 21–33.

TACITUS (A.D. 98). 'A Treatise on the Situation, Manners and Inhabitants of Germany.'
    Bohn's Classical Library, vol. ii.

G. H. THOMSON (1919). 'The Criterion of Goodness of Fit of Psychophysical Curves.'
    Biometrika, xii, p. 216.

H. H. TURNER (1924). 'On the Numerical Aspect of Reciprocal Mimicry (Diapose-
    matic Resemblance). *Trans. Ent. Soc.* (1924), pp. 667–75.

A. R. WALLACE (1889). *Darwinism.* London: Macmillan.

E. A. WESTERMARCK (1906). *Origin and Development of Moral Ideas.* Chapter, XVII.

W. M. WHEELER (1923). *Social Life among the Insects*, p. 225. London: Constable.

W. C. D. WHETHAM (1909). *The Family and the Nation*, p. 140.

O. WINGE (1923). 'Crossing-over between the X- and Y-chromosome in *Lebistes*.'
    Comptes Rendus des travaux du laboratoire de Carlsberg, vol. xiv, No. 20.

SEWALL WRIGHT (1925). 'The Factors of the Albino Series of Guinea Pigs and their
    Effects on Black and Yellow Pigmentation.' *Genetics*, vol. x, pp. 223–60.

INDEX

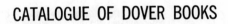
CATALOGUE OF DOVER BOOKS

Biological Sciences

AN INTRODUCTION TO GENETICS, A. H. Sturtevant and G. W. Beadle. A very thorough exposition of genetic analysis and the chromosome mechanics of higher organisms by two of the world's most renowned biologists, A. H. Sturtevant, one of the founders of modern genetics, and George Beadle, Nobel laureate in 1958. Does not concentrate on the biochemical approach, but rather more on observed data from experimental evidence and results . . . from Drosophila and other life forms. Some chapter titles: Sex chromosomes; Sex-Linkage; Autosomal Inheritance;; Chromosome Maps; Intra-Chromosomal Rearrangements; Inversions—and Incomplete Chromosomes; Translocations; Lethals; Mutations; Heterogeneous Populations; Genes and Phenotypes; The Determination and Differentiation of Sex; etc. Slightly corrected reprint of 1939 edition. New preface by Drs. Sturtevant and Beadle. 1 color plate. 126 figures. Bibliographies. Index. 391pp. 5⅜ x 8½.   S306 Paperbound **$2.00**

THE GENETICAL THEORY OF NATURAL SELECTION, R. A. Fisher. 2nd revised edition of a vital reviewing of Darwin's Selection Theory in terms of particulate inheritance, by one of the great authorities on experimental and theoretical genetics. Theory is stated in mathematical form. Special features of particulate inheritance are examined: evolution of dominance, maintenance of specific variability, mimicry and sexual selection, etc. 5 chapters on man and his special circumstances as a social animal. 16 photographs. Bibliography. Index. x + 310pp. 5⅜ x 8.   S466 Paperbound **$2.00**

THE ORIENTATION OF ANIMALS: KINESES, TAXES AND COMPASS REACTIONS, Gottfried S. Fraenkel and Donald L. Gunn. A basic work in the field of animal orientations. Complete, detailed survey of everything known in the subject up to 1940s, enlarged and revised to cover major developments to 1960. Analyses of simpler types of orientation are presented in Part I: kinesis, klinotaxis, tropotaxis, telotaxis, etc. Part II covers more complex reactions originating from temperature changes, gravity, chemical stimulation, etc. The two-light experiment and unilateral blinding are dealt with, as is the problem of determinism or volition in lower animals. The book has become the universally-accepted guide to all who deal with the subject—zoologists, biologists, psychologists, and the like. Second, enlarged edition, revised to 1960. Bibliography of over 500 items. 135 illustrations. Indices. xiii + 376pp. 5⅜ x 8½.   T786 Paperbound **$2.25**

THE BEHAVIOUR AND SOCIAL LIFE OF HONEYBEES, C. R. Ribbands. Definitive survey of all aspects of honeybee life and behavior; completely scientific in approach, but written in interesting, everyday language that both professionals and laymen will appreciate. Basic coverage of physiology, anatomy, sensory equipment; thorough account of honeybee behavior in the field (foraging activities, nectar and pollen gathering, how individuals find their way home and back to food areas, mating habits, etc.); details of communication in various field and hive situations. An extensive treatment of activities within the hive community—food sharing, wax production, comb building, swarming, the queen, her life and relationship with the workers, etc. A must for the beekeeper, natural historian, biologist, entomologist, social scientist, et al. "An indispensable reference," J. Hambleton, BtES. "Recommended in the strongest of terms," AMERICAN SCIENTIST. 9 plates. 66 figures. Indices. 693-item bibliography. 252pp. 5⅜ x 8½.   T1137 Paperbound **$2.00**

BIRD DISPLAY: AN INTRODUCTION TO THE STUDY OF BIRD PSYCHOLOGY, E. A. Armstrong. The standard work on bird display, based on extensive observation by the author and reports of other observers. This important contribution to comparative psychology covers the behavior and ceremonial rituals of hundreds of birds from gannet and heron to birds of paradise and king penguins. Chapters discuss such topics as the ceremonial of the gannet, ceremonial gaping, disablement reactions, the expression of emotions, the evolution and function of social ceremonies, social hierarchy in bird life, dances of birds and men, songs, etc. Free of technical terminology, this work will be equally interesting to psychologists and zoologists as well as bird lovers of all backgrounds. 32 photographic plates. New introduction by the author. List of scientific names of birds. Bibliography. 3-part index. 431pp. 5⅜ x 8½.   T1128 Paperbound **$2.00**

THE SPECIFICITY OF SEROLOGICAL REACTIONS, Karl Landsteiner. With a Chapter on Molecular Structure and Intermolecular Forces by Linus Pauling. Dr. Landsteiner, winner of the Nobel Prize in 1930 for the discovery of the human blood groups, devoted his life to fundamental research and played a leading role in the development of immunology. This authoritative study is an account of the experiments he and his colleagues carried out on antigens and serological reactions with simple compounds. Comprehensive coverage of the basic concepts of immunolgy includes such topics as: The Serological Specificity of Proteins, Antigens, Antibodies, Artificially Conjugated Antigens, Non-Protein Cell Substances such as polysaccharides, etc., Antigen-Antibody Reactions (Toxin Neutralization, Precipitin Reactions, Agglutination, etc.). Discussions of toxins, bacterial proteins, viruses, hormones, enzymes, etc. in the context of immunological phenomena. New introduction by Dr. Merrill Chase of the Rockefeller Institute. Extensive bibliography and bibliography of author's writings. Index. xviii + 330pp. 5⅜ x 8½.   S299 Paperbound **$2.00**

CATALOGUE OF DOVER BOOKS

CULTURE METHODS FOR INVERTEBRATE ANIMALS, P. S. Galtsoff, F. E. Lutz, P. S. Welch, J. G. Needham, eds. A compendium of practical experience of hundreds of scientists and technicians, covering invertebrates from protozoa to chordata, in 313 articles on 17 phyla. Explains in great detail food, protection, environment, reproduction conditions, rearing methods, embryology, breeding seasons, schedule of development, much more. Includes at least one species of each considerable group. Half the articles are on class insecta. Introduction. 97 illustrations. Bibliography. Index. xxix + 590pp. 5⅜ x 8. S526 Paperbound **$3.00**

THE BIOLOGY OF THE LABORATORY MOUSE, edited by G. D. Snell. 1st prepared in 1941 by the staff of the Roscoe B. Jackson Memorial Laboratory, this is still the standard work on the mouse, assembling an enormous amount of material for which otherwise you spend hours of research. Embryology, reproduction, histology, spontaneous tumor formation, genetics of tumor transplantation, endocrine secretion & tumor formation, milk, influence & tumor formation, inbred, hybrid animals, parasites, infectious diseases, care & recording. Classified bibliography of 1122 items. 172 figures, including 128 photos. ix + 497pp. 6⅛ x 9¼.
S248 Clothbound **$6.00**

MATHEMATICAL BIOPHYSICS: PHYSICO-MATHEMATICAL FOUNDATIONS OF BIOLOGY, N. Rashevsky. One of most important books in modern biology, now revised, expanded with new chapters, to include most significant recent contributions. Vol. 1: Diffusion phenomena, particularly diffusion drag forces, their effects. Old theory of cell division based on diffusion drag forces, other theoretical approaches, more exhaustively treated than ever. Theories of excitation, conduction in nerves, with formal theories plus physico-chemical theory. Vol. 2: Mathematical theories of various phenomena in central nervous system. New chapters on theory of color vision, of random nets. Principle of optimal design, extended from earlier edition. Principle of relational mapping of organisms, numerous applications. Introduces into mathematical biology such branches of math as topology, theory of sets. Index. 236 illustrations. Total of 988pp. 5⅜ x 8. S574 Vol. 1 (Books 1, 2) Paperbound **$2.50**
S575 Vol. 2 (Books 3, 4) Paperbound **$2.50**
2 vol. set **$5.00**

ELEMENTS OF MATHEMATICAL BIOLOGY, A. J. Lotka. A pioneer classic, the first major attempt to apply modern mathematical techniques on a large scale to phenomena of biology, biochemistry, psychology, ecology, similar life sciences. Partial Contents: Statistical meaning of irreversibility; Evolution as redistribution; Equations of kinetics of evolving systems; Chemical, inter-species equilibrium; parameters of state; Energy transformers of nature, etc. Can be read with profit even by those having no advanced math; unsurpassed as study-reference. Formerly titled ELEMENTS OF PHYSICAL BIOLOGY. 72 figures. xxx + 460pp. 5⅜ x 8. S346 Paperbound **$2.45**

THE BIOLOGY OF THE AMPHIBIA, G. K. Noble, Late Curator of Herpetology at the Am. Mus. of Nat. Hist. Probably the most used text on amphibia, unmatched in comprehensiveness, clarity, detail. 19 chapters plus 85-page supplement cover development; heredity; life history; speciation; adaptation; sex, integument, respiratory, circulatory, digestive, muscular, nervous systems; instinct, intelligence, habits, environment, economic value, relationships, classification, etc. "Nothing comparable to it," C. H. Pope, Curator of Amphibia, Chicago Mus. of Nat. Hist. 1047 bibliographic references. 174 illustrations. 600pp. 5⅜ x 8.
S206 Paperbound **$2.98**

STUDIES ON THE STRUCTURE AND DEVELOPMENT OF VERTEBRATES, E. S. Goodrich. A definitive study by the greatest modern comparative anatomist. Exceptional in its accounts of the ossicles of the ear, the separate divisions of the coelom and mammalian diaphragm, and the 5 chapters devoted to the head region. Also exhaustive morphological and phylogenetic expositions of skeleton, fins and limbs, skeletal visceral arches and labial cartilages, visceral clefts and gills, vacular, respiratory, excretory, and peripheral nervous systems, etc., from fish to the higher mammals. 754 illustrations. 69 page biographical study by C. C. Hardy. Bibliography of 1186 references. "What an undertaking . . . to write a textbook which will summarize adequately and succinctly all that has been done in the realm of Vertebrate Morphology these recent years," Journal of Anatomy. Index. Two volumes. Total 906pp. 5⅜ x 8. Two vol. set S449-50 Paperbound **$5.00**

A TREATISE ON PHYSIOLOGICAL OPTICS, H. von Helmholtz, Ed. by J. P. C. Southall. Unmatched for thoroughness, soundness, and comprehensiveness, this is still the most important work ever produced in the field of physiological optics. Revised and annotated, it contains everything known about the subject up to 1925. Beginning with a careful anatomical description of the eye, the main body of the text is divided into three general categories: The Dioptrics of the Eye (covering optical imagery, blur circles on the retina, the mechanism of accommodation, chromatic aberration, etc.); The Sensations of Vision (including stimulation of the organ of vision, simple and compound colors, the intensity and duration of light, variations of sensitivity, contrast, etc.); and The Perceptions of Vision (containing movements of the eyes, the monocular field of vision, direction, perception of depth, binocular double vision, etc.). Appendices cover later findings on optical imagery, refraction, ophthalmoscopy, and many other matters. Unabridged, corrected republication of the original English translation of the third German edition. 3 volumes bound as 2. Complete bibliography, 1911-1925. Indices. 312 illustrations. 6 full-page plates, 3 in color. Total of 1,749pp. 5⅜ x 8.
Two-volume set S15, 16 Clothbound **$15.00**

CATALOGUE OF DOVER BOOKS

INTRODUCTION TO PHYSIOLOGICAL OPTICS, James P. C. Southall, former Professor of Physics in Columbia University. Readable, top-flight introduction, not only for beginning students of optics, but also for other readers—physicists, biochemists, illuminating engineers, optometrists, psychologists, etc. Comprehensive coverage of such matters as the Organ of Vision (structure of the eyeball, the retina, the dioptric system, monocular and binocular vision, adaptation, etc.); The Optical System of the Eye (reflex images in the cornea and crystalline lens, Emmetropia and Ametropia, accommodation, blur circles on retina); Eye-Glasses; Eye Defects; Movements of the Eyeball in its Socket; Rod and Cone Vision; Color Vision; and other similar topics. Index. 134 figures. x +426pp. 5⅜ x 8.   S924 Paperbound **$2.25**

LIGHT, COLOUR AND VISION, Yves LeGrand. A thorough examination of the eye as a receptor of radiant energy and as a mechanism (the retina) consisting of light-sensitive cells which absorb light of various wave lengths—probably the most complete and authoritative treatment of this subject in print. Originally prepared as a series of lectures given at the Institute of Optics in Paris, subsequently enlarged for book publication. Partial contents: Radiant Energy—concept, nature, theories, etc., Sources of Radiation—artificial and natural, the Visual Receptor, Photometric Quantities, Units, Calculations, Retinal Illumination, Trivariance of Vision, Colorimetry, Luminance Difference Thresholds, Anatomy of the Retina, Theories of Vision, Photochemistry and Electro-physiology of the Retina, etc. Appendices, Exercises, with solutions. 500-item bibliography. Authorized translation by R. Hunt, J. Walsh, F. Hunt. Index. 173 illustrations. xiii + 512pp. 5⅜ x 8½.   S979 Clothbound **$10.00**

FINGER PRINTS, PALMS AND SOLES: AN INTRODUCTION TO DERMATOGLYPHICS, Harold Cummins and Charles Midlo. An introduction in non-technical language designed to acquaint the reader with a long-neglected aspect of human biology. Although a chapter dealing with fingerprint identification and the systems of classification used by the FBI, etc. has been added especially for this edition, the main concern of the book is to show how the intricate pattern of ridges and wrinkles on our fingers have a broader significance, applicable in many areas of science and life. Some topics are: the identification of two types of twins; the resolution of doubtful cases of paternity; racial variation; inheritance; the relation of fingerprints to body measurements, blood groups, criminality, character, etc. Classification and recognition of fundamental patterns and pattern types discussed fully. 149 figures. 49 tables. 361-item bibliography. Index. xii + 319pp. 5⅝ x 8⅜.   T778 Paperbound **$2.25**

Classics and histories

ANTONY VAN LEEUWENHOEK AND HIS "LITTLE ANIMALS," edited by Clifford Dobell. First book to treat extensively, accurately, life and works (relating to protozoology, bacteriology) of first microbiologist, bacteriologist, micrologist. Includes founding papers of protozoology, bacteriology; history of Leeuwenhoek's life; discussions of his microscopes, methods, language. His writing conveys sense of an enthusiastic, naive genius, as he looks at rainwater, pepper water, vinegar, frog's skin, rotifers, etc. Extremely readable, even for non-specialists. "One of the most interesting and enlightening books I have ever read," Dr. C. C. Bass, former Dean, Tulane U. School of Medicine. Only authorized edition. 400-item bibliography. Index. 32 illust. 442pp. 5⅜ x 8.   S594 Paperbound **$2.25**

THE GROWTH OF SCIENTIFIC PHYSIOLOGY, G. J. Goodfield. A compact, superbly written account of how certain scientific investigations brought about the emergence of the distinct science of physiology. Centers principally around the mechanist-vitalist controversy prior to the development of physiology as an independent science, using the arguments which raged around the problem of animal heat as its chief illustration. Covers thoroughly the efforts of clinicians and naturalists and workers in chemistry and physics to solve these problems—from which the new discipline arose. Includes the theories and contributions of: Aristotle, Galen, Harvey, Boyle, Bernard, Benjamin Franklin, Palmer, Gay-Lussac, Priestley, Spallanzani, and many others. 1960 publication. Biographical bibliography. 174pp. 5 x 7½.   T1066 Clothbound **$3.00**

MICROGRAPHIA, Robert Hooke. Hooke, 17th century British universal scientific genius, was a major pioneer in celestial mechanics, optics, gravity, and many other fields, but his greatest contribution was this book, now reprinted entirely from the original 1665 edition, which gave microscopy its first great impetus. With all the freshness of discovery, he describes fully his microscope, and his observations of cork, the edge of a razor, insects' eyes, fabrics, and dozens of other different objects. 38 plates, full-size or larger, contain all the original illustrations. This book is also a fundamental classic in the fields of combustion and heat theory, light and color theory, botany and zoology, hygrometry, and many other fields. It contains such farsighted predictions as the famous anticipation of artificial silk. The final section is concerned with Hooke's telescopic observations of the moon and stars. 323pp. 5⅜ x 8.   T8 Paperbound **$2.50**

Medicine

CLASSICS OF MEDICINE AND SURGERY, edited by C. N. B. Camac. 12 greatest papers in medical history, 11 in full: Lister's "Antiseptic Principle;" Harvey's "Motion in the Heart and Blood;" Auenbrugger's "Percussion of the Chest;" Laënnec's "Auscultation and the Stethoscope;" Jenner's "Inquiry into Smallpox Vaccine," 2 related papers; Morton's "Administering Sulphuric Ether," letters to Warren, "Physiology of Ether;" Simpson's "A New Anaesthetic Agent;" Holmes' "Puerperal Fever." Biographies, portraits of authors, bibliographies. Formerly "Epoch-making Contributions to Medicine, Surgery, and the Allied Sciences." Introduction. 14 illus. 445pp. 5⅜ x 8. S539 Paperbound **$2.25**

A WAY OF LIFE, Sir William Osler. The complete essay, stating his philosophy of life, as given at Yale University by this great physician and teacher. 30 pages. Copies limited, no more than 1 to a customer. Free.

SOURCE BOOK OF MEDICAL HISTORY, compiled, annotated by Logan Clendening, M.D. Unequalled collection of 139 greatest papers in medical history, by 120 authors, covers almost every area: pathology, asepsis, preventive medicine, bacteriology, physiology, etc. Hippocrates, Gain, Vesalius, Malpighi, Morgagni, Boerhave, Pasteur, Walter Reed, Florence Nightingale, Lavoisier, Claude Bernard, 109 others, give view of medicine unequalled for immediacy. Careful selections give heart of each paper save you reading time. Selections from non-medical literature show lay-views of medicine: Aristophanes, Plato, Arabian Nights, Chaucer, Molière, Dickens, Thackeray, others. "Notable . . . useful to teacher and student alike," Amer. Historical Review. Bibliography. Index. 699pp. 5⅜ x 8. T621 Paperbound **$2.75**

EXPERIMENTS AND OBSERVATIONS ON THE GASTRIC JUICE AND THE PHYSIOLOGY OF DIGESTION, William Beaumont. A gunshot wound which left a man with a 2½ inch hole through his abdomen into his stomach (1822) enabled Beaumont to perform the remarkable experiments set down here. The first comprehensive, thorough study of motions and processes of the stomach, "his work remains a model of patient, persevering investigation. . . . Beaumont is the pioneer physiologist of this country." (Sir William Osler, in his introduction.) 4 illustrations. xi + 280pp. 5⅜ x 8. S527 Paperbound **$1.50**

AN INTRODUCTION TO THE STUDY OF EXPERIMENTAL MEDICINE, Claude Bernard. 90-year-old classic of medical science, only major work of Bernard available in English, records his efforts to transform physiology into exact science. Principles of scientific research illustrated by specific case histories from his work; roles of chance, error, preliminary false conclusions, in leading eventually to scientific truth; use of hypothesis. Much of modern application of mathematics to biology rests on the foundation set down here. New foreword by Professor I. B. Cohen, Harvard Univ. xxv + 266pp. 5⅜ x 8. T400 Paperbound **$1.50**

A WAY OF LIFE, AND OTHER SELECTED WRITINGS, Sir William Osler, Physician and humanist, Osler discourses brilliantly in thought provoking essays and on the history of medicine. He discusses Thomas Browne, Gui Patin, Robert Burton, Michael Servetus, William Beaumont, Laënnec. Includes such favorite writings as the title essay, "The Old Humanities and the New Science," "Creators, Transmitters, and Transmuters," "Books and Men," "The Student Life," and five more of his best discussions of philosophy, religion and literature. 5 photographs. Introduction by G. L. Keynes, M.D., F.R.C.S. Index. xx + 278pp. 5⅜ x 8. T488 Paperbound **$1.50**

THE HISTORY OF SURGICAL ANESTHESIA, Thomas E. Keys. Concise, but thorough and always engrossing account of the long struggle to find effective methods of eliminating pain during surgery, tracing the remarkable story through the centuries to the eventual successes by dedicated researchers, the acceptance of ether, the work of men such as Priestley, Morton, Lundy, and many, many others. Discussions of the developments in local, regional, and spinal anesthesia, etc. "The general reader as well as the medical historian will find material to interest him in this fascinating story," U.S. QUARTERLY BOOKLIST. Revised, enlarged publication of original edition. Introductory essay by C. D. Leake. Concluding chapter by N. A. Gillespie. Appendix by J. F. Fulton. 46 illustrations. New preface by the author. Chronology of events. Extensive bibliographies. Index. xxx + 193pp. 5⅜ x 8½. T1122 Paperbound **$2.00**

A SHORT HISTORY OF ANATOMY AND PHYSIOLOGY FROM THE GREEKS TO HARVEY, Charles Singer. Corrected edition of THE EVOLUTION OF ANATOMY, classic work tracing evolution of anatomy and physiology from prescientific times through Greek & Roman periods, Dark Ages, Renaissance, to age of Harvey and beginning of modern concepts. Centered on individuals, movements, periods that definitely advanced anatomical knowledge: Plato, Diocles, Aristotle, Theophrastus, Herophilus, Erasistratus, the Alexandrians, Galen, Mondino, da Vinci, Linacre, Sylvius, others. Special section on Vesalius; Vesalian atlas of nudes, skeletons, muscle tabulae. Index of names, 20 plates. 270 extremely interesting illustrations of ancient, medieval, Renaissance, Oriental origin. xii + 209pp. 5⅜ x 8. T389 Paperbound **$1.75**

Books Explaining Science and Mathematics

WHAT IS SCIENCE?, N. Campbell. The role of experiment and measurement, the function of mathematics, the nature of scientific laws, the difference between laws and theories, the limitations of science, and many similarly provocative topics are treated clearly and without technicalities by an eminent scientist. "Still an excellent introduction to scientific philosophy," H. Margenau in PHYSICS TODAY. "A first-rate primer . . . deserves a wide audience," SCIENTIFIC AMERICAN. 192pp. 5⅜ x 8. S43 Paperbound **$1.25**

THE NATURE OF PHYSICAL THEORY, P. W. Bridgman. A Nobel Laureate's clear, non-technical lectures on difficulties and paradoxes connected with frontier research on the physical sciences. Concerned with such central concepts as thought, logic, mathematics, relativity, probability, wave mechanics, etc. he analyzes the contributions of such men as Newton, Einstein, Bohr, Heisenberg, and many others. "Lucid and entertaining . . . recommended to anyone who wants to get some insight into current philosophies of science," THE NEW PHILOSOPHY. Index. xi + 138pp. 5⅜ x 8. S33 Paperbound **$1.25**

EXPERIMENT AND THEORY IN PHYSICS, Max Born. A Nobel Laureate examines the nature of experiment and theory in theoretical physics and analyzes the advances made by the great physicists of our day: Heisenberg, Einstein, Bohr, Planck, Dirac, and others. The actual process of creation is detailed step-by-step by one who participated. A fine examination of the scientific method at work. 44pp. 5⅜ x 8. S308 Paperbound **75¢**

THE PSYCHOLOGY OF INVENTION IN THE MATHEMATICAL FIELD, J. Hadamard. The reports of such men as Descartes, Pascal, Einstein, Poincaré, and others are considered in this investigation of the method of idea-creation in mathematics and other sciences and the thinking process in general. How do ideas originate? What is the role of the unconscious? What is Poincaré's forgetting hypothesis? are some of the fascinating questions treated. A penetrating analysis of Einstein's thought processes concludes the book. xiii + 145pp. 5⅜ x 8. T107 Paperbound **$1.25**

THE NATURE OF LIGHT AND COLOUR IN THE OPEN AIR, M. Minnaert. Why are shadows sometimes blue, sometimes green, or other colors depending on the light and surroundings? What causes mirages? Why do multiple suns and moons appear in the sky? Professor Minnaert explains these unusual phenomena and hundreds of others in simple, easy-to-understand terms based on optical laws and the properties of light and color. No mathematics is required but artists, scientists, students, and everyone fascinated by these "tricks" of nature will find thousands of useful and amazing pieces of information. Hundreds of observational experiments are suggested which require no special equipment. 200 illustrations; 42 photos. xvi + 362pp. 5⅜ x 8. T196 Paperbound **$2.00**

***MATHEMATICS IN ACTION, O. G. Sutton.** Everyone with a command of high school algebra will find this book one of the finest possible introductions to the application of mathematics to physical theory. Ballistics, numerical analysis, waves and wavelike phenomena, Fourier series, group concepts, fluid flow and aerodynamics, statistical measures, and meteorology are discussed with unusual clarity. Some calculus and differential equations theory is developed by the author for the reader's help in the more difficult sections. 88 figures. Index. viii + 236pp. 5⅜ x 8. T440 Clothbound **$3.50**

SOAP-BUBBLES: THEIR COLOURS AND THE FORCES THAT MOULD THEM, C. V. Boys. For continuing popularity and validity as scientific primer, few books can match this volume of easily-followed experiments, explanations. Lucid exposition of complexities of liquid films, surface tension and related phenomena, bubbles' reaction to heat, motion, music, magnetic fields. Experiments with capillary attraction, soap bubbles on frames, composite bubbles, liquid cylinders and jets, bubbles other than soap, etc. Wonderful introduction to scientific method, natural laws that have many ramifications in areas of modern physics. Only complete edition in print. New Introduction by S. Z. Lewin, New York University. 83 illustrations; 1 full-page color plate. xii + 190pp. 5⅜ x 8½. T542 Paperbound **95¢**

History of Science and Mathematics

THE STUDY OF THE HISTORY OF MATHEMATICS, THE STUDY OF THE HISTORY OF SCIENCE, G. Sarton. Two books bound as one. Each volume contains a long introduction to the methods and philosophy of each of these historical fields, covering the skills and sympathies of the historian, concepts of history of science, psychology of idea-creation, and the purpose of history of science. Prof. Sarton also provides more than 80 pages of classified bibliography. Complete and unabridged. Indexed. 10 illustrations. 188pp. 5⅜ x 8.   T240 Paperbound **$1.25**

A HISTORY OF PHYSICS, Florian Cajori, Ph.D. First written in 1899, thoroughly revised in 1929, this is still best entry into antecedents of modern theories. Precise non-mathematical discussion of ideas, theories, techniques, apparatus of each period from Greeks to 1920's, analyzing within each period basic topics of matter, mechanics, light, electricity and magnetism, sound, atomic theory, etc. Stress on modern developments, from early 19th century to present. Written with critical eye on historical development, significance. Provides most of needed historical background for student of physics. Reprint of second (1929) edition. Index. Bibliography in footnotes. 16 figures. xv + 424pp. 5⅜ x 8.   T970 Paperbound **$2.00**

A HISTORY OF ASTRONOMY FROM THALES TO KEPLER, J. L. E. Dreyer. Formerly titled A HISTORY OF PLANETARY SYSTEMS FROM THALES TO KEPLER. This is the only work in English which provides a detailed history of man's cosmological views from prehistoric times up through the Renaissance. It covers Egypt, Babylonia, early Greece, Alexandria, the Middle Ages, Copernicus, Tycho Brahe, Kepler, and many others. Epicycles and other complex theories of positional astronomy are explained in terms nearly everyone will find clear and easy to understand. "Standard reference on Greek astronomy and the Copernican revolution," SKY AND TELESCOPE. Bibliography. 21 diagrams. Index. xvii + 430pp. 5⅜ x 8.   S79 Paperbound **$2.25**

A SHORT HISTORY OF ASTRONOMY, A. Berry. A popular standard work for over 50 years, this thorough and accurate volume covers the science from primitive times to the end of the 19th century. After the Greeks and Middle Ages, individual chapters analyze Copernicus, Brahe, Galileo, Kepler, and Newton, and the mixed reception of their startling discoveries. Post-Newtonian achievements are then discussed in unusual detail: Halley, Bradley, Lagrange, Laplace, Herschel, Bessel, etc. 2 indexes. 104 illustrations, 9 portraits. xxxi + 440pp. 5⅜ x 8.   T210 Paperbound **$2.00**

PIONEERS OF SCIENCE, Sir Oliver Lodge. An authoritative, yet elementary history of science by a leading scientist and expositor. Concentrating on individuals—Copernicus, Brahe, Kepler, Galileo, Descartes, Newton, Laplace, Herschel, Lord Kelvin, and other scientists—the author presents their discoveries in historical order, adding biographical material on each man and full, specific explanations of their achievements. The full, clear discussions of the accomplishments of post-Newtonian astronomers are features seldom found in other books on the subject. Index. 120 illustrations. xv + 404pp. 5⅜ x 8.   T716 Paperbound **$1.65**

THE BIRTH AND DEVELOPMENT OF THE GEOLOGICAL SCIENCES, F. D. Adams. The most complete and thorough history of the earth sciences in print. Geological thought from earliest recorded times to the end of the 19th century—covers over 300 early thinkers and systems: fossils and hypothetical explanations of them, vulcanists vs. neptunists, figured stones and paleontology, generation of stones, and similar topics. 91 illustrations, including medieval, renaissance woodcuts, etc. 632 footnotes and bibliographic notes. Index. 511pp. 5⅜ x 8.   T5 Paperbound **$2.25**

THE STORY OF ALCHEMY AND EARLY CHEMISTRY, J. M. Stillman. "Add the blood of a red-haired man"—a recipe typical of the many quoted in this authoritative and readable history of the strange beliefs and practices of the alchemists. Concise studies of every leading figure in alchemy and early chemistry through Lavoisier, in this curious epic of superstition and true science, constructed from scores of rare and difficult Greek, Latin, German, and French texts. Foreword by S. W. Young. 246-item bibliography. Index. xiii + 566pp. 5⅜ x 8.   S628 Paperbound **$2.45**

HISTORY OF MATHEMATICS, D. E. Smith. Most comprehensive non-technical history of math in English. Discusses the lives and works of over a thousand major and minor figures, from Euclid to Descartes, Gauss, and Riemann. Vol. I: A chronological examination, from primitive concepts through Egypt, Babylonia, Greece, the Orient, Rome, the Middle Ages, the Renaissance, and up to 1900. Vol. 2: The development of ideas in specific fields and problems, up through elementary calculus. Two volumes, total of 510 illustrations, 1355pp. 5⅜ x 8. Set boxed in attractive container.   T429,430 Paperbound the set **$5.00**

Classics of Science

THE DIDEROT PICTORIAL ENCYCLOPEDIA OF TRADES AND INDUSTRY, MANUFACTURING AND THE TECHNICAL ARTS IN PLATES SELECTED FROM "L'ENCYCLOPEDIE OU DICTIONNAIRE RAISONNE DES SCIENCES, DES ARTS, ET DES METIERS" OF DENIS DIDEROT, edited with text by C. Gillispie. The first modern selection of plates from the high point of 18th century French engraving, Diderot's famous Encyclopedia. Over 2000 illustrations on 485 full page plates, most of them original size, illustrating the trades and industries of one of the most fascinating periods of modern history, 18th century France. These magnificent engravings provide an invaluable glimpse into the past for the student of early technology, a lively and accurate social document to students of cultures, an outstanding find to the lover of fine engravings. The plates teem with life, with men, women, and children performing all of the thousands of operations necessary to the trades before and during the early stages of the industrial revolution. Plates are in sequence, and show general operations, closeups of difficult operations, and details of complex machinery. Such important and interesting trades and industries are illustrated as sowing, harvesting, beekeeping, cheesemaking, operating windmills, milling flour, charcoal burning, tobacco processing, indigo, fishing, arts of war, salt extraction, mining, smelting iron, casting iron, steel, extracting mercury, zinc, sulphur, copper, etc., slating, tinning, silverplating, gilding, making gunpowder, cannons, bells, shoeing horses, tanning, papermaking, printing, dying, and more than 40 other categories. 920pp. 9 x 12. Heavy library cloth.   T421 Two volume set **$18.50**

THE PRINCIPLES OF SCIENCE, A TREATISE ON LOGIC AND THE SCIENTIFIC METHOD, W. Stanley Jevons. Treating such topics as Inductive and Deductive Logic, the Theory of Number, Probability, and the Limits of Scientific Method, this milestone in the development of symbolic logic remains a stimulating contribution to the investigation of inferential validity in the natural and social sciences. It significantly advances Boole's logic, and describes a machine which is a foundation of modern electronic calculators. In his introduction, Ernest Nagel of Columbia University says, "(Jevons) . . . continues to be of interest as an attempt to articulate the logic of scientific inquiry." Index. liii + 786pp. 5⅜ x 8.
S446 Paperbound **$2.98**

*DIALOGUES CONCERNING TWO NEW SCIENCES, Galileo Galilei. A classic of experimental science which has had a profound and enduring influence on the entire history of mechanics and engineering. Galileo based this, his finest work, on 30 years of experimentation. It offers a fascinating and vivid exposition of dynamics, elasticity, sound, ballistics, strength of materials, and the scientific method. Translated by H. Crew and A. de Salvio. 126 diagrams. Index. xxi + 288pp. 5⅜ x 8.   S99 Paperbound **$1.75**

DE MAGNETE, William Gilbert. This classic work on magnetism founded a new science. Gilbert was the first to use the word "electricity," to recognize mass as distinct from weight, to discover the effect of heat on magnetic bodies; invented an electroscope, differentiated between static electricity and magnetism, conceived of the earth as a magnet. Written by the first great experimental scientist, this lively work is valuable not only as an historical landmark, but as the delightfully easy-to-follow record of a perpetually searching, ingenious mind. Translated by P. F. Mottelay. 25 page biographical memoir. 90 fix. lix + 368pp. 5⅜ x 8.   S470 Paperbound **$2.00**

*OPTICKS, Sir Isaac Newton. An enormous storehouse of insights and discoveries on light, reflection, color, refraction, theories of wave and corpuscular propagation of light, optical apparatus, and mathematical devices which have recently been reevaluated in terms of modern physics and placed in the top-most ranks of Newton's work! Foreword by Albert Einstein. Preface by I. B. Cohen of Harvard U. 7 pages of portraits, facsimile pages, letters, etc. cxvi + 412pp. 5⅜ x 8.   S205 Paperbound **$2.25**

A SURVEY OF PHYSICAL THEORY, M. Planck. Lucid essays on modern physics for the general reader by the Nobel Laureate and creator of the quantum revolution. Planck explains how the new concepts came into being; explores the clash between theories of mechanics, electrodynamics, and thermodynamics; and traces the evolution of the concept of light through Newton, Huygens, Maxwell, and his own quantum theory, providing unparalleled insights into his development of this momentous modern concept. Bibliography. Index. vii + 121pp. 5⅜ x 8.
S650 Paperbound **$1.15**

A SOURCE BOOK IN MATHEMATICS, D. E. Smith. English translations of the original papers that announced the great discoveries in mathematics from the Renaissance to the end of the 19th century: succinct selections from 125 different treatises and articles, most of them unavailable elsewhere in English—Newton, Leibniz, Pascal, Riemann, Bernoulli, etc. 24 articles trace developments in the field of number, 18 cover algebra, 36 are on geometry, and 13 on calculus. Biographical-historical introductions to each article. Two volume set. Index in each. Total of 115 illustrations. Total of xxviii + 742pp. 5⅜ x 8.
S552 Vol I Paperbound **$2.00**
S553 Vol II Paperbound **$2.00**
The set, boxed **$4.00**

CATALOGUE OF DOVER BOOKS

*THE THIRTEEN BOOKS OF EUCLID'S ELEMENTS, edited by T. L. Heath.** This is the complete EUCLID — the definitive edition of one of the greatest classics of the western world. Complete English translation of the Heiberg text with spurious Book XIV. Detailed 150-page introduction discusses aspects of Greek and medieval mathematics: Euclid, texts, commentators, etc. Paralleling the text is an elaborate critical exposition analyzing each definition, proposition, postulate, etc., and covering textual matters, mathematical analyses, refutations, extensions, etc. Unabridged reproduction of the Cambridge 2nd edition. 3 volumes. Total of 995 figures, 1426pp. 5⅜ x 8.   S88, 89, 90 — 3 vol. set, Paperbound **$7.50**

*THE GEOMETRY OF RENE DESCARTES.** The great work which founded analytic geometry. The renowned Smith-Latham translation faced with the original French text containing all of Descartes' own diagrams! Contains: Problems the Construction of Which Requires Only Straight Lines and Circles; On the Nature of Curved Lines; On the Construction of Solid or Supersolid Problems. Notes. Diagrams. 258pp.   S68 Paperbound **$1.60**

*A PHILOSOPHICAL ESSAY ON PROBABILITIES, P. Laplace.** Without recourse to any mathematics above grammar school, Laplace develops a philosophically, mathematically and historically classical exposition of the nature of probability: its functions and limitations, operations in practical affairs, calculations in games of chance, insurance, government, astronomy, and countless other fields. New introduction by E. T. Bell. viii + 196pp.   S166 Paperbound **$1.35**

DE RE METALLICA, Georgius Agricola. Written over 400 years ago, for 200 years the most authoritative first-hand account of the production of metals, translated in 1912 by former President Herbert Hoover and his wife, and today still one of the most beautiful and fascinating volumes ever produced in the history of science! 12 books, exhaustively annotated, give a wonderfully lucid and vivid picture of the history of mining, selection of sites, types of deposits, excavating pits, sinking shafts, ventilating, pumps; crushing machinery, assaying, smelting, refining metals, making salt, alum, nitre, glass, and many other topics. This definitive edition contains all 289 of the 16th century woodcuts which made the original an artistic masterpiece. It makes a superb gift for geologists, engineers, libraries, artists, historians, and everyone interested in science and early illustrative art. Biographical, historical introductions. Bibliography, survey of ancient authors. Indices. 289 illustrations. 672pp. 6¾ x 10¾. Deluxe library edition.   S6 Clothbound **$10.00**

GEOGRAPHICAL ESSAYS, W. M. Davis. Modern geography and geomorphology rest on the fundamental work of this scientist. His new concepts of earth-processes revolutionized science and his broad interpretation of the scope of geography created a deeper understanding of the interrelation of the landscape and the forces that mold it. This first inexpensive unabridged edition covers theory of geography, methods of advanced geographic teaching, descriptions of geographic areas, analyses of land-shaping processes, and much besides. Not only a factual and historical classic, it is still widely read for its reflections of modern scientific thought. Introduction. 130 figures. Index. vi + 777pp. 5⅜ x 8.
S383 Paperbound **$3.50**

CHARLES BABBAGE AND HIS CALCULATING ENGINES, edited by P. Morrison and E. Morrison. Friend of Darwin, Humboldt, and Laplace, Babbage was a leading pioneer in large-scale mathematical machines and a prophetic herald of modern operational research—true father of Harvard's relay computer Mark I. His Difference Engine and Analytical Engine were the first successful machines in the field. This volume contains a valuable introduction on his life and work; major excerpts from his fascinating autobiography, revealing his eccentric and unusual personality; and extensive selections from "Babbage's Calculating Engines," a compilation of hard-to-find journal articles, both by Babbage and by such eminent contributors as the Countess of Lovelace, L. F. Menabrea, and Dionysius Lardner. 11 illustrations. Appendix of miscellaneous papers. Index. Bibliography. xxxviii + 400pp. 5⅜ x 8.   T12 Paperbound **$2.25**

*THE WORKS OF ARCHIMEDES WITH THE METHOD OF ARCHIMEDES, edited by T. L. Heath.** All the known works of the greatest mathematician of antiquity including the recently discovered METHOD OF ARCHIMEDES. This last is the only work we have which shows exactly how early mathematicians discovered their proofs before setting them down in their final perfection. A 186 page study by the eminent scholar Heath discusses Archimedes and the history of Greek mathematics. Bibliography. 563pp. 5⅜ x 8.   S9 Paperbound **$2.45**

Psychology

YOGA: A SCIENTIFIC EVALUATION, Kovoor T. Behanan. A complete reprinting of the book that for the first time gave Western readers a sane, scientific explanation and analysis of yoga. The author draws on controlled laboratory experiments and personal records of a year as a disciple of a yoga, to investigate yoga psychology, concepts of knowledge, physiology, "supernatural" phenomena, and the ability to tap the deepest human powers. In this study under the auspices of Yale University Institute of Human Relations, the strictest principles of physiological and psychological inquiry are followed throughout. Foreword by W. A. Miles, Yale University. 17 photographs. Glossary. Index. xx + 270pp. 5⅜ x 8. T505 Paperbound **$2.00**

CONDITIONED REFLEXES: AN INVESTIGATION OF THE PHYSIOLOGICAL ACTIVITIES OF THE CEREBRAL CORTEX, I. P. Pavlov. Full, authorized translation of Pavlov's own survey of his work in experimental psychology reviews entire course of experiments, summarizes conclusions, outlines psychological system based on famous "conditioned reflex" concept. Details of technical means used in experiments, observations on formation of conditioned reflexes, function of cerebral hemispheres, results of damage, nature of sleep, typology of nervous system, significance of experiments for human psychology. Trans. by Dr. G. V. Anrep, Cambridge Univ. 235-item bibliography. 18 figures. 445pp. 5⅜ x 8. S614 Paperbound **$2.35**

EXPLANATION OF HUMAN BEHAVIOUR, F. V. Smith. A major intermediate-level introduction to and criticism of 8 complete systems of the psychology of human behavior, with unusual emphasis on theory of investigation and methodology. Part I is an illuminating analysis of the problems involved in the explanation of observed phenomena, and the differing viewpoints on the nature of causality. Parts II and III are a closely detailed survey of the systems of McDougall, Gordon Allport, Lewin, the Gestalt group, Freud, Watson, Hull, and Tolman. Biographical notes. Bibliography of over 800 items. 2 indexes. 38 figures. xii + 460pp. 5½ x 8¾. T253 Clothbound **$6.00**

SEX IN PSYCHO-ANALYSIS (formerly CONTRIBUTIONS TO PSYCHO-ANALYSIS), S. Ferenczi. Written by an associate of Freud, this volume presents countless insights on such topics as impotence, transference, analysis and children, dreams, symbols, obscene words, masturbation and male homosexuality, paranoia and psycho-analysis, the sense of reality, hypnotism and therapy, and many others. Also includes full text of THE DEVELOPMENT OF PSYCHO-ANALYSIS by Ferenczi and Otto Rank. Two books bound as one. Total of 406pp. 5⅜ x 8. T324 Paperbound **$1.85**

BEYOND PSYCHOLOGY, Otto Rank. One of Rank's most mature contributions, focussing on the irrational basis of human behavior as a basic fact of our lives. The psychoanalytic techniques of myth analysis trace to their source the ultimates of human existence: fear of death, personality, the social organization, the need for love and creativity, etc. Dr. Rank finds them stemming from a common irrational source, man's fear of final destruction. A seminal work in modern psychology, this work sheds light on areas ranging from the concept of immortal soul to the sources of state power. 291pp. 5⅜ x 8. T485 Paperbound **$2.00**

ILLUSIONS AND DELUSIONS OF THE SUPERNATURAL AND THE OCCULT, D. H. Rawcliffe. Holds up to rational examination hundreds of persistent delusions including crystal gazing, automatic writing, table turning, mediumistic trances, mental healing, stigmata, lycanthropy, live burial, the Indian Rope Trick, spiritualism, dowsing, telepathy, clairvoyance, ghosts, ESP, etc. The author explains and exposes the mental and physical deceptions involved, making this not only an exposé of supernatural phenomena, but a valuable exposition of characteristic types of abnormal psychology. Originally titled "The Psychology of the Occult." 14 illustrations. Index. 551pp. 5⅜ x 8. T503 Paperbound **$2.00**

THE PRINCIPLES OF PSYCHOLOGY, William James. The full long-course, unabridged, of one of the great classics of Western literature and science. Wonderfully lucid descriptions of human mental activity, the stream of thought, consciousness, time perception, memory, imagination, emotions, reason, abnormal phenomena, and similar topics. Original contributions are integrated with the work of such men as Berkeley, Binet, Mills, Darwin, Hume, Kant, Royce, Schopenhauer, Spinoza, Locke, Descartes, Galton, Wundt, Lotze, Herbart, Fechner, and scores of others. All contrasting interpretations of mental phenomena are examined in detail — introspective analysis, philosophical interpretation, and experimental research. "A classic," JOURNAL OF CONSULTING PSYCHOLOGY. "The main lines are as valid as ever," PSYCHO-ANALYTICAL QUARTERLY. "Standard reading . . . a classic of interpretation," PSYCHIATRIC QUARTERLY. 94 illustrations. 1408pp. 2 volumes. 5⅜ x 8. Vol. 1, T381 Paperbound **$2.50**
Vol. 2, T382 Paperbound **$2.50**

THE DYNAMICS OF THERAPY IN A CONTROLLED RELATIONSHIP, Jessie Taft. One of the most important works in literature of child psychology, out of print for 25 years. Outstanding disciple of Rank describes all aspects of relationship or Rankian therapy through concise, simple elucidation of theory underlying her actual contacts with two seven-year olds. Therapists, social caseworkers, psychologists, counselors, and laymen who work with children will all find this important work an invaluable summation of method, theory of child psychology. xix + 296pp. 5⅜ x 8. T325 Paperbound **$1.75**

SELECTED PAPERS ON HUMAN FACTORS IN THE DESIGN AND USE OF CONTROL SYSTEMS, Edited by H. Wallace Sinaiko. Nine of the most important papers in this area of increasing interest and rapid growth. All design engineers who have encountered problems involving man as a system-component will find this volume indispensable, both for its detailed information about man's unique capacities and defects, and for its comprehensive bibliography of articles and journals in the human-factors field. Contributors include Chapanis, Birmingham, Adams, Fitts and Jones, etc. on such topics as Theory and Methods for Analyzing Errors in Man-Machine Systems, A Design Philosophy for Man-Machine Control Systems, Man's Senses as Informational Channels, The Measurement of Human Performance, Analysis of Factors Contributing to 460 "Pilot Error" Experiences, etc. Name, subject indexes. Bibliographies of over 400 items. 27 figures. 8 tables. ix + 405pp. 6⅛ x 9¼.   S140 Paperbound **$2.75**

THE ANALYSIS OF SENSATIONS, Ernst Mach. Great study of physiology, psychology of perception, shows Mach's ability to see material freshly, his "incorruptible skepticism and independence." (Einstein). Relation of problems of psychological perception to classical physics, supposed dualism of physical and mental, principle of continuity, evolution of senses, will as organic manifestation, scores of experiments, observations in optics, acoustics, music, graphics, etc. New introduction by T. S. Szasz, M. D. 58 illus. 300-item bibliography. Index. 404pp. 5⅜ x 8.   S525 Paperbound **$1.75**

PRINCIPLES OF ANIMAL PSYCHOLOGY, N. R. F. Maier and T. C. Schneirla. The definitive treatment of the development of animal behavior and the comparative psychology of all animals. This edition, corrected by the authors and with a supplement containing 5 of their most important subsequent articles, is a "must" for biologists, psychologists, zoologists, and others. First part of book includes analyses and comparisons of the behavior of characteristic types of animal life—from simple multicellular animals through the evolutionary scale to reptiles and birds, tracing the development of complexity in adaptation. Two-thirds of the book covers mammalian life, developing further the principles arrived at in Part I. New preface by the authors. 153 illustrations and tables. Extensive bibliographic material. Revised indices. xvi + 683pp. 5⅜ x 8½.   S1120 Paperbound **$3.00** (tentative)

ERROR AND ECCENTRICITY IN HUMAN BELIEF, Joseph Jastrow. From 180 A.D. to the 1930's, the surprising record of human credulity: witchcraft, miracle workings, animal magnetism, mind-reading, astral-chemistry, dowsing, numerology, etc. The stories and exposures of the theosophy of Madame Blavatsky and her followers, the spiritism of Helene Smith, the imposture of Kaspar Hauser, the history of the Ouija board, the puppets of Dr. Luy, and dozens of other hoaxers and cranks, past and present. "As a potpourri of strange beliefs and ideas, it makes excellent reading," New York Times. Formerly titled "Wish and Wisdom, Episodes in the Vagaries of Belief." Unabridged publication. 56 illustrations and photos. 22 full-page plates. Index. xv + 394pp. 5⅜ x 8½.   T986 Paperbound **$1.85**

THE PHYSICAL DIMENSIONS OF CONSCIOUSNESS, Edwin G. Boring. By one of the ranking psychologists of this century, a major work which reflected the logical outcome of a progressive trend in psychological theory—a movement away from dualism toward physicalism. Boring, in this book, salvaged the most important work of the structuralists and helped direct the mainstream of American psychology into the neo-behavioristic channels of today. Unabridged republication of original (1933) edition. New preface by the author. Indexes. 17 illustrations. xviii + 251pp. 5⅜ x 8.   S1040 Paperbound **$1.75**

BRAIN MECHANISMS AND INTELLIGENCE: A QUANTITATIVE STUDY OF INJURIES TO THE BRAIN, K. S. Lashley. A major contemporary psychologist examines the influence of brain injuries upon the capacity to learn, retentiveness, the formation of the maze habit, etc. Also: the relation of reduced learning ability to sensory and motor defects, the nature of the deterioration following cerebral lesions, comparison of the rat with other forms, and related matters. New introduction by Prof. D. O. Hebb. Bibliography. Index. xxii + 200pp. 5⅜ x 8½.   T1038 Paperbound **$1.75**

Prices subject to change without notice.

Dover publishes books on art, music, philosophy, literature, languages, history, social sciences, psychology, handcrafts, orientalia, puzzles and entertainments, chess, pets and gardens, books explaining science, intermediate and higher mathematics, mathematical physics, engineering, biological sciences, earth sciences, classics of science, etc. Write to:

Dept. catrr.
Dover Publications, Inc.
180 Varick Street, N.Y. 14, N.Y.